Living on ONE ACRE OR LESS

To the memory of my grandfather, George Wilkes,
who let me have my own small plot in the garden
and sowed the seed of my love of plants.

Living on ONE ACRE OR LESS

How to produce all the fruit, vegetables, meat, fish and eggs your family needs

Sally Morgan

Published by
Green Books
An imprint of UIT Cambridge Ltd
www.greenbooks.co.uk

PO Box 145, Cambridge CB4 1GQ, England
+44 (0)1223 302 041

First published in 2016, in England

All interior images are by Sally Morgan, with the exception of the following: Page 49: Ecoscene / Aldridge. Page 52 (left): Ecoscene / Gryniewicz. Page 120: Ecoscene / Ian Harwood. Page 136 (bottom): Ecoscene / Kathryn Martin. Page 156: Pam Drake / Cornwall Turkeys. Page 171: Ecoscene / Mark Tweedie. Page 190: FAO Aquaculture Photo Library. Page 191: Ryan Griffis. Page 202: Ecoscene / Chris Gill. Pages 55 & 145: iStock. Pages 30, 144, 148, 175, 181, 204, 208 & 195 (x 4): Shutterstock. The image on page 189 shows the Aquaponics Solar Greenhouse at Humble by Nature. The illustrations on pages 193 & 196 are based on reference material from The Aquaponic Source.

Design by Jayne Jones

ISBN: 978-0-85784-330-2 (paperback)
ISBN: 978-0-85784-331-9 (ePub)
ISBN: 978-0-85784-332-6 (pdf)
Also available for Kindle

10 9 8 7 6 5 4 3 2 1

CONTENTS

A small but productive permaculture garden in Hong Kong.

INTRODUCTION

Many people dream of stepping away from the rat race and living on a plot of land where they can be self-sufficient: growing all their own fruit and vegetables, keeping chickens for eggs, and perhaps raising a few pigs or sheep. But with the soaring cost of land and the difficulty of finding the right plot, the reality is that many will probably have to make do with a large garden or a corner of a field. The good news is that it is perfectly possible to grow all the fruit and vegetables needed by a family, raise animals for meat, fish and eggs, keep bees and even produce fuelwood on a small plot of land of just one acre or less.

Much of the developing world relies on the productivity of small farms. There are more than 500 million smallholder farms around the world, and more than 2 billion people rely on them for their livelihoods. Between them, these small farms produce four-fifths of all the food eaten in Asia and sub-Saharan Africa. More than 50 years ago, the Nobel-Prize-winning economist Amartya Sen showed that there was an inverse relationship between the size of a farm and the amount of crops it could produce per hectare: the smaller the farm, the greater the productivity. A recent study compared farm size and yield across seven regions in Turkey, finding the smaller farms to be many times more productive than larger farms, and this seems to be true everywhere, even in the UK and USA.

There are plenty of reasons to explain this fact. While some have suggested that small-scale farmers are able to make use of free family labour to cultivate more intensively, in fact there is little doubt that the productivity actually comes from biodiverse systems with an abundance of wildlife, the use of polycultures (growing different crops together) and a fertile soil. Most use organic methods, which can be as productive as conventional methods, especially on a small scale. For example, long-term experiments at the Rodale Institute in the USA have found that yields of organically grown crops match conventionally grown yields in an average year, and can even outperform them in drought years. Organic methods build up fertility in the soil, use less energy and are far more sustainable.

Many small-scale farmers are self-reliant, growing everything they need, from food and fibres to fuel and medicines. I say 'self-reliant' deliberately, since it is not quite the same thing as self-sufficient, as I am frequently reminded by permaculturalist friends! It is virtually impossible to be truly self-sufficient, but being self-reliant means that you have goods that you can trade for the items that you cannot produce.

Another word that crops up a lot in this context is 'resilience' – the ability to recover from difficult

The small plots on these terraced slopes in Madeira are intensively worked to provide families with all the vegetables they need.

conditions; to be adaptable and flexible. Small-scale farming is far more resilient than large-scale commercial farming. Small-scale farmers tend to grow a diverse range of crops rather than specialize in one or two, and can therefore adapt to suit changing conditions and recover quickly from adverse conditions. In a world that is experiencing climate change, resilience is going to be very important.

I never fail to be impressed by the productivity of smallholders and gardeners around the world, having seen some amazing examples – from terraced slopes in Madeira and Hong Kong to tiny plots in Kenya; from vegetable gardens in Australia and North America to allotments in the UK. There is something to be learned from all of them. This book is a compilation of all the nuggets of information that I have gleaned on my journey, from my early days of growing vegetables with my grandparents as a child, then studying botany at university, carrying out research into habitat restoration and, more recently, my experiences of keeping livestock and running a smallholding. I don't think farmers and growers ever stop learning. Today, I use ideas from permaculture and organic horticulture, and I love trying new ways based on experimental work that I have seen at research stations and from my travels worldwide.

In this book I hope to show you just how much can be produced in a small area, using traditional techniques rather than 'quick fix' chemical approaches. My advice to people taking on land for the first time is always to take it slowly. Too many people rush off and buy lots of different animals, create a huge vegetable-growing area, and then find that it becomes a chore rather than a pleasure, which often leads to them

Highly productive and organically managed terraces at the Kadoorie Farm in Hong Kong.

giving up. Hopefully, the advice in this book will help you to avoid this. Some of the information you will need to put together your plans is to be found in Parts Two and Three, so my best advice is to read everything before starting!

How to use this book

This book is divided into three parts. Part One covers the process of getting started, helping you to survey and design your plot, put up fencing and build paths. It looks at the soil too, which is so essential to a productive plot, and I explain how to build soil fertility through composting and growing green manures. Part Two focuses on growing fruit, vegetables and flowers, establishing trees for fuel and even developing a forest garden. In Part Three I turn to the broad subject of livestock, and discuss the keeping of poultry, pigs, sheep and goats, as well as fish and bees.

Throughout this book I have provided measurements in both metric and imperial units and referred to seasons in generic terms rather than by specific months of the year, in order to make the text as versatile as possible, relevant to readers in any location.

The rules and regulations regarding the keeping of livestock vary from country to country, so I have touched upon this area only briefly within the main chapters, but you will find more details in the Appendix. If you do decide to keep animals such as pigs, sheep and goats, and I hope you do, please check with the relevant local authorities regarding the regulations that need to be followed.

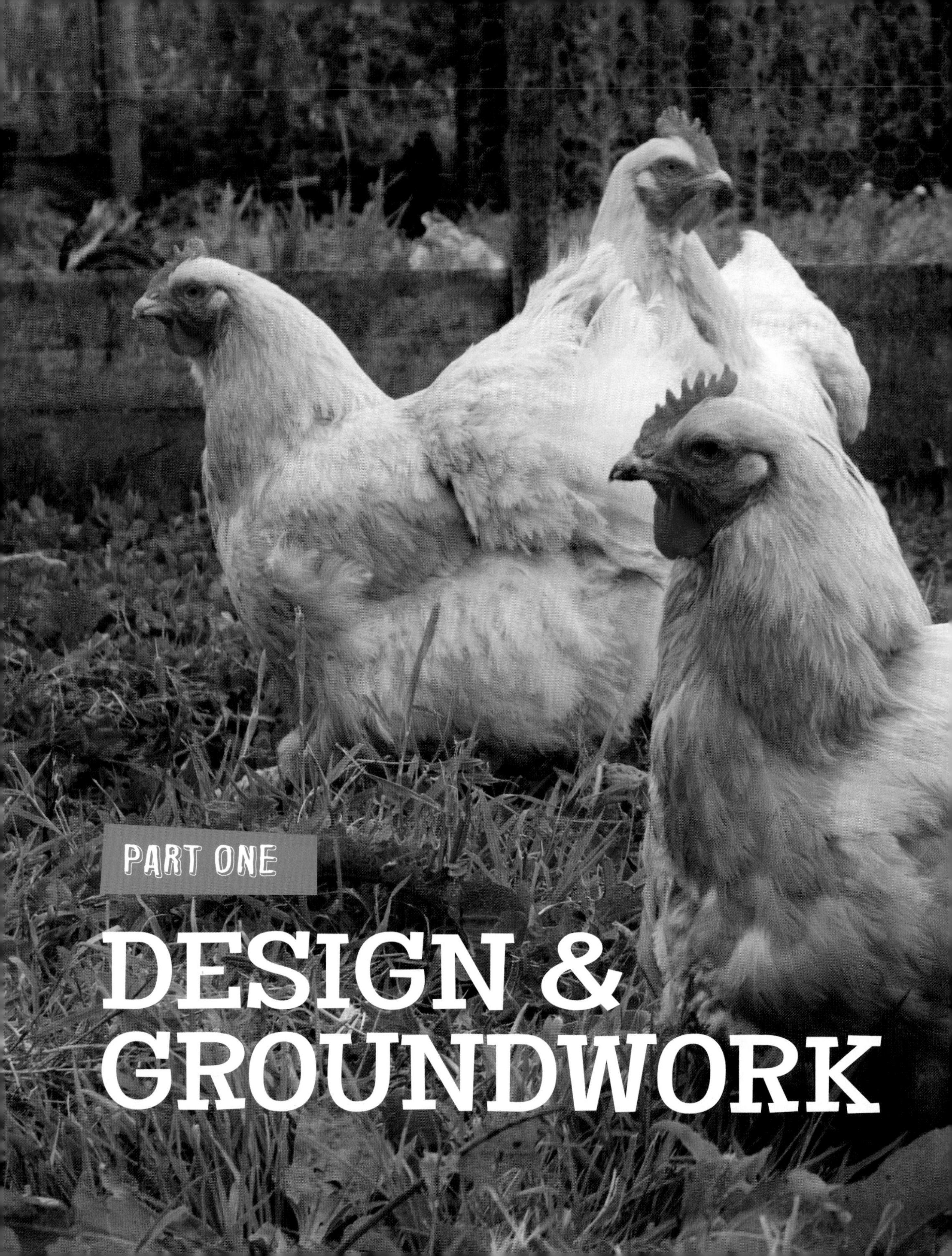

PART ONE

DESIGN & GROUNDWORK

Chapter ONE

Planning your plot

Farming on a small plot is intensive, but in a good way. On even less than an acre you can keep poultry, a couple of pigs, some bees and fish. If you have enough grass, you can raise some lambs or keep goats. And there will still be room for fruit and vegetable plots, sheds and polytunnels.

Our walled garden is just two-thirds of an acre, but there is plenty of space to squeeze in a small orchard.

Many people are surprised when they stand on our one-acre plot and see its extent: when asked, they estimate it to be much larger. They realize that an acre is a lot of space and that even a slightly smaller plot will keep them busy! In this chapter, I'll take you through the first steps in setting up your plot: of planning and design, erecting fences and building paths, and putting up polytunnels (which are essential for productivity in a temperate climate). I'll also tackle the topic of pest control.

The good thing about having a small plot is that it ensures that you make use of all the available space, through methods such as growing under cover, using raised beds and growing vertically. However, success is dependent on a good-quality soil that is rich in organic matter and nutrients, to supply plants with all their needs, and this must be backed up by an effective composting system. With good soil fertility, it is possible to really boost crop yields. For example, as much as 50kg (110lb) of carrots can be harvested from as little as $3m^2$ (32 sq ft) if the plants are closely planted in a raised bed filled with fertile soil – a yield many times that which is possible in a commercial situation. But care has to be taken that the system doesn't become too intensive, as it is very easy to plant too many crops, and that is simply not sustainable. You have to keep to a reasonable level of production, or you will have to start relying on fertilizers and other chemicals because the balance has been lost.

Similarly, when working with a small plot of land, do not be tempted to keep too many animals. You need to make sure you have a good land-use rotation to maintain the quality of the soil and minimize the burden of parasites and disease organisms in your animals, and if you have too many animals this is not feasible. Wherever possible, your animals should be part of your crop rotation to make best use of the free supply of nutrients in their dung (see Chapter 3).

When planning your plot, it is critical to be realistic about just how much time you have to spend. If you are not living on-site and are unable to visit it on a daily basis, then opt for something that won't take up much of your time. The vegetable area, for example, could occupy half the plot, but you would have to put in many hours a week just to keep control of it – and what will you do with all the produce? Sell it or give it away? Realistically, you could manage with much less for your own veg supply, and have an orchard for fruit or a small paddock for pigs or lamb.

With a polytunnel you could focus on niche crops such as chillies.

Running a business

An acre can be enough to set up a viable small commercial enterprise. For example, you could run a niche business such as an organic vegetable box scheme, supplying seasonal vegetables, fruit, eggs and honey to as many as 60 to 70 customers. You could have a small orchard, soft fruit and a herb garden, and make seasonal preserves and fruit juices. A good-sized area under polytunnels would allow you to grow chillies and other exotics.

Your vegetable area could occupy half the plot, but you would have to put in many hours a week just to keep control of it.

Planning on paper

Whether you have acquired a greenfield plot or taken over an existing smallholding, it is almost certain that you will need to make changes. It's always best to construct a plan on paper before tackling any work on the site, so you can make decisions on the position of fences, paths and structures such as polytunnels. First, make a scale drawing of your plot showing the boundaries and any existing buildings, and key features such as hedgerows, trees, fences, and water and electric points. Add a compass 'rose' and delineate the shady areas, and indicate the direction of prevailing winds.

Make accurate measurements of the plot before you start. You can do this on the ground, but nowadays it's quicker and easier from the comfort of your desk by using a GPS area measurement app on your smartphone that makes use of Google Earth maps. I checked the accuracy of one of these apps and found it to be accurate to within 0.5m (1'8"), which is perfectly adequate for a plan. There are apps that will measure the height of objects – another useful feature if you have trees, buildings or a tall hedge nearby. It is useful to know the height of objects such as trees, as you can then work out the shelter effect. A windbreak is effective for up to six times its height, so a 3m (10') hedge on the prevailing wind side of your plot will give 18m (about 60') of shelter. Drones, too, can be useful: you can use one to take an aerial photo of the site, as shown in the photos on the next two pages.

Planning checklist

The following are key points to consider when making your plan.

Decide on an access point if there is not an obvious one already (check the planning regulations); the size of gates and whether you need a hard-standing or turning area.

Decide on the type of boundary you need (fence or hedge), and whether you will need windbreaks (rows of willow, poplar or Jerusalem artichokes, for example).

A slope will restrict what you can do, and you'll need to plan carefully to avoid soil erosion, with terraces across the slope rather than exposed areas on the slope. You may want to dig swales (ditches) along the contour lines to slow down and capture water run-off. Before you do anything, go out during heavy rain and observe how the water runs off your land.

Think about the size and position of the more permanent structures such as sheds and polytunnels, which will need good access, will create shade and will generate some water run-off.

Consider security carefully, especially if your site is visible from a road or footpath.

Where will be your main routes through the plot? These will be formed between the various points that you need to access, and the type of route may depend on the frequency of use (see also 'Zoning', below). How many paths will you need and how wide do they have to be?

Where will your water come from? Are you going to have pipes and taps or will you collect water from roofs? Do you need a dipping pond or a water tank?

Where will you build your vegetable beds? They will need an open, sunny location, which should be flat if possible. The best orientation

A drone fitted with a camera can provide a new perspective of your plot.

An overhead of our one-acre plot, with polytunnel, vegetable beds and livestock pens.

Once you have a rough plan, check that you have made best use of the available space. Could you add an edible hedgerow, or cordon fruit trees along the boundaries?

for the beds is north–south, so all plants get the same amount of sunlight and do not shade each other. If the orientation is east–west, then there will be some shading.

Mark out the area that you will use for permanent crops, including fruit trees and soft fruits. Do you want an orchard, or will you plant fruit and nut trees across the plot? Do you have space for a fruit cage?

Where will you build your compost bins? You will need space to wheel in the green waste and turn the compost. Don't forget that you will be barrowing the finished compost on to the vegetable beds, so it needs to be close to them and not down a slope!

Do you want to grow willow or poplar for fuel or to include a forest garden?

Make a list of all the animals you want to keep, and work out the area of land you need for them. Remember that you may need trailer access for transporting pigs, bringing in chicken houses or moving fencing materials, etc. Read the advice in Part Three before making these decisions.

Once you have a rough plan, check that you have made best use of all the space. Can you add an edible hedgerow, or plant cordon or espalier fruit trees along the boundaries? Have you collected all the water that may be available to you? Do you have enough space for animals and options for rotating their pasture?

Before you get carried away and start planting trees and buying livestock, it is absolutely critical to get the site layout right, because paths, fences and buildings are expensive to put in and can't be moved easily. The first step is to plan your boundary fencing if there is none in place or if what is there is unsuitable; this is particularly important if you want to keep livestock. You'll need to decide what type of paths you want and their routes, and the location and type of any large enclosures and any structures such as polytunnels and sheds, before going on to site smaller things such as your planting areas and compost heaps. Check the local planning rules regarding sheds and polytunnels.

Zoning

When designing your plot, it is useful to think of 'zones': a permaculture principle based on the number of visits an area will receive on a daily basis. If you will be visiting an area very frequently, it makes sense to locate it near to your access point (if you are living on-site then your access point in this sense will be your house). Broadly speaking, you can imagine different zones as concentric circles or sectors, spreading out from the access point. However, zones should not be defined rigidly, as there may be

A glade in a forest garden creates a variety of microclimates and supports more biodiversity.

The edge effect

It is well known that in nature the most biodiverse and productive areas are those in the overlap zone between two ecosystems: for example, in the intertidal zone along coasts, where the open ocean meets land; in mangrove forests along tropical coastlines; where forest meets grassland; and in woodland clearings. This is because in these areas there is a greater range of conditions, allowing more species of plants to survive, which in turn supports a greater diversity of animals. If you can incorporate more 'edges' in your plot, you will create lots of different microclimates and a richer ecosystem. For example, I like to have small log piles, where beneficial predators such as beetles and spiders can hide, and I let the grass grow long in places as this attracts different animals. If you are planning a forest garden (see Chapter 6), include a small glade or clearing that aligns north to south. It will have its own microclimate, being sheltered from extremes by the trees all around it, with (in the northern hemisphere) the northern end getting plenty of sunlight from the south. Another way of creating more edge is simply to have wavy edges rather than straight lines between areas. A curved path has more edge than a straight path, while a keyhole or a spiral bed are more varied in microclimates than a rectangular bed.

A water feature

A small pond on the plot will attract useful animals, such as frogs and toads and possibly grass snakes, and provide drinking water for mammals such as hedgehogs. It doesn't have to be large, and could perhaps be sited in a quiet corner where the wildlife won't be disturbed. If you are harvesting water on the site, any overflow water could be directed into the pond. In our walled garden, water from one of our barns is directed into a dipping pond, and from there any overflow goes into a large duck and carp pond (see Chapter 10). The water in the dipping pond is enough to water the vegetable beds all summer, and it's quick and easy to fill a bucket with water, so a hose is unnecessary.

When designing your plot, it's useful to think in terms of the permaculture concept of **zones.**

other constraints on your site, such as slope and aspect, that determine where something will be sited, and you can also have a zone appearing more than once on your plan.

Zone 1 describes those areas that need the most frequent visits, in order to water, tend and harvest, so these are generally positioned closest to the access point. They might include the polytunnel, cold frames, shed, aquaculture unit, and vegetable beds and wormeries.

Zone 2 will be areas that receive less frequent daily attention, such as the chicken and pig pens, where the animals still need checking twice a day but are less intensively managed than in zone 1.

Zone 3 is the areas that will not require daily visits, such as the orchard, fruit cage, perennial beds, composting bins and crops that need less attention, such as squash; plus your bee hives. You might include your sheep and goats in this zone, although they will require two checks a day.

Zone 4 might include your forest garden, willow for fuel plot, and carp ponds.

Zone 5 will be your wild areas that are rarely visited, so are often located near the furthest boundaries.

Fencing

There are various factors to consider when deciding on the location of a fenced pasture. The shape makes a big difference to the length of fence required for a given area: the squarer it is, the less fence you need. Avoid any acute corners, as they are awkward to get into with machinery. Shade, slope, access to water and the type of livestock are also key factors. The priority must be fences along roads and around

Positioning posts for stock fences

Setting posts correctly is one of the most important factors in fence strength. The correct depth depends on the diameter of the post and your soil type. Generally, in medium to heavy clay soils, a post is placed at a depth equal to 10 times its diameter. In sandier soils, the depth should be 15 times the diameter. At these depths, the post would break before it uproots. The spacing of posts varies with the fence type and the lie of the land. Posts are usually positioned every 3m (10'), but where the land slopes, posts are placed at the top and bottom of the slope to ensure the fence follows the contours.

areas from which livestock must be excluded or contained: these will normally be fixed or permanent fences. Paddocks can be subdivided with temporary fences, such as flexi-nets (see page 24) to change the grazing areas and to respond to changes in grass availability.

Stock fencing

Stock fencing is a long-term, permanent option, and (as the name suggests) is used to contain livestock. It comes in a range of sizes, with different heights and mesh patterns to suit different animals, and can be used with two additional lines of wire to increase the height of the fence. The netting is fixed to either round or half-round stakes and is supported at ends, at corners and at every change of direction by straining posts, which have bracing struts for additional strength. If you are fencing in pigs, which take great delight in digging under unprotected fencing, it is best to add a single strand of electric fencing along the bottom on the inside.

Electric fencing

Electric fencing is the quickest and most cost-effective way to contain livestock. It is easy to install and repair, and usually requires fewer posts than stock fencing. Electric fence can be temporary or permanent. You will need reels of wire or tape or netting, posts, insulators, an energizer, and a power source such as a 12V battery or access to mains power. Such electric fences require no tools for set-up, minimal bracing, and use lightweight plastic line posts.

Electric fences, whether portable or permanent, use galvanized wire, polywire or polytape to transmit the shock. Polywire is a braided or twisted polyethylene cord with three, six or nine strands of stainless-steel wire running through it. Polytape, as the name suggests, is a woven flat polyethylene tape, also with several wires running through it. It is more visible than polywire, and it tends to flutter in the wind, making it easy for animals to spot and avoid. It is the wire that carries the electricity in both polytape and polywire, so the more strands the better: nine is best, six is adequate, but avoid three. I prefer to use galvanized wire because, even though it is a little harder to handle, it is much more effective electrically (it is easier to get a good join and has less electrical loss).

It is almost always best to use galvanized wire for permanent electric fencing and for fencing around pigs, though we often add a polytape

Our permanent 9-strand electric fence around our poultry field has lasted for 10 years with minimal maintenance.

Electric fencing is the quickest and most cost-effective way to contain livestock, and is easy to install and repair.

top line so that animals and humans can easily see the fence (See Chapter 8, page 169). Due to the long lengths of wire, reels are an essential part of putting out and retrieving electric fences. If fences are to be erected and dismantled regularly, it may pay to invest in a geared reel.

Permanent electric fences will need either wood or steel corner posts with insulators, though you can still use plastic intermediate line posts. Whether wood or plastic, the line post needs to be rigid enough to withstand wind but flexible enough to bend under excessive pressure. It is important to keep the fence free of vegetation, so you can either strim under the fence regularly or, to avoid this necessity, place a strip of ground-cover fabric along the line of the fence. At gateways you can put the wire in a length of piping, so people can step over it safely, or use sprung coils of wire to create a removable barrier.

These hens are contained by 50m (164') of electric poultry netting, which is 110cm (3'8") tall.

TIP Invest in a fence checker

When you have a busy routine, it's easy to forget to check that your electric fence is working every day - and you certainly can't afford for your animals to get out or a fox to get in. So invest in a fence checker. These are hung on one of the wires in an obvious position; ours flashes red if there is a problem, but others work by showing a green light when all is working correctly.

Remember that you should put warning signs on the fence if it is somewhere the general public could come into contact with it, such as beside a footpath.

Flexi-nets

One option for poultry and sheep is flexible electric netting. It usually comes complete with plastic posts and a mains- or battery-powered battery unit, and can be moved from place to place quite easily and rolled up when not needed. It comes in different heights and is generally sold in lengths of 25m or 50m (82' or 164') with gates. A 50m roll creates a pen of approximately 156m^2 (1,680 sq ft). The difficulty with this form of netting, however, is keeping the bottom strands free of vegetation. Although it

Solar panels can recharge energizer batteries automatically. They must be sited in direct sunlight and oriented to receive the maximum amount of light possible.

can be used as an external perimeter for sheep, it is better as an internal barrier or divider in a stock-fenced field.

Energizers

It is important to choose a suitable energizer for the fence. It can be powered by mains electricity or by 6V or 12V DC batteries; some use solar panels to recharge the batteries. Energizers are often rated in terms of length of wire: the longer the fence, the more powerful the energizer has to be in order to send an effective charge throughout its length. All energizers need to cope with adverse conditions, such as wet weather, contact with vegetation, bad ground conditions and poor joins in the wire. The voltage present at any point on the fence where an animal comes into contact with it typically ranges from 2,000V to 4,000V. Usually 2,000V is sufficient for cattle, but 4,000V is better for sheep and poultry, due to the animals' naturally thick insulation.

Grounding

One of the most common failures in an electric fencing system is poor grounding, which results in weak shocks. The electricity must be able to complete a circuit back to the energizer through the ground, so you have to fit ground rods of copper or steel to provide an effective 'earth'.

Paths

Access to the various parts of your plot is so much easier if you get the position of the paths right. But remember that paths can take up valuable growing space, so think carefully before positioning them. You will need a wide path or track with gentle turns for easy access to the paddocks for your livestock trailer and other large vehicles. Footpaths should be wide enough to take a wheelbarrow, so 60-70cm (24-28"); while the narrowest paths can be 30cm (12"), just enough to let you walk along and squat down.

On our one-acre plot, we have an unfenced grassy area in front of the pens. This is around 5m (16') wide, and we have access through a double gate for a tractor and trailer, so we can load up the pigs when they go to slaughter, or move sheep if necessary. Also, using an electric flexi-net, we can create a temporary enclosure for a chicken run or for grazing a couple of lambs or a gaggle of geese.

Remember that paths can take up valuable growing space, so think carefully before positioning them.

Laying paths and tracks is an expensive business, so, where possible, opt for grassy paths which can be kept in check by regularly mowing in summer. For those areas where you don't want grass or where grass simply won't stand up to the traffic, you will probably want hogging (a compacted cover of gravel, sand and clay) or gravel. An intermediate option is to simply lay down a strip of ground-cover fabric. Depending on the soil type, you could simply leave the soil bare, if it's not likely to get too muddy, as hoeing and walking will stop weeds germinating.

In areas where you will have heavy vehicles parking and turning, it pays to prepare the area carefully. Dig out the surface to a depth of about 15-20cm (6-8"). Construct an edge with gravel boards and then lay a sub-base of crushed stone, rubble or hardcore, which is hammered down with a vibrating-plate compactor. To get a solid base without any gaps, you need to have a good mix of fine and large particles. The surface is then dressed with gravel. This type of track can last for many years.

A well-laid gravel path will last many years, needing just the occasional top-up with gravel.

Our polytunnel lies in an east-west orientation near the entrance to the plot.

Polytunnels

An essential element of any holding in a temperate climate is an unheated polytunnel or large greenhouse, for extending the cropping season. Polytunnels are not the most attractive of structures, but they are much cheaper and easier to erect than greenhouses, and are therefore a good option unless there is a greenhouse already on-site or you can adapt an existing building. A polytunnel generally needs less maintenance than a greenhouse, although the plastic will need to be replaced every 4 to 7 years.

I spend many hours in our polytunnel throughout the year. It is a large one, but I have no difficulty filling it with all manner of crops. There is space to put out the essentials, such as summer tomatoes, chillies, cucumbers, melons and salad crops, and still room to experiment with more unusual crops such as achocha or okra. We have crops growing in the polytunnel all year round; it's a real pleasure to be able to eat freshly harvested leeks, spinach and celeriac in the middle of winter!

Our polytunnel is 25m by 9m (82' by 30') and is oriented east–west. There is a hedge on the west side, far enough away to provide some shelter from the prevailing winds but not close enough to create shade on the tunnel. The hoops are buried in concrete for support and the plastic is buried in the ground.

This polytunnel is covered with a laminated plastic, which is much stronger than the usual type of polythene sheet.

Choosing and siting your polytunnel

Before deciding on the type and position of a polytunnel, there are various points to consider.

Orientation Most people opt for an east–west orientation, which means there is sun along the length of the tunnel all day, with a hot side and a cooler side. But temperatures inside during the hottest months can reach unbearable levels, whereas a north–south orientation, with its restricted exposure to the sun during the hottest hours of the day, means lower temperatures are experienced.

Site levels Obviously, it's better to erect your polytunnel on a flat site, as it makes installation much easier. It is possible to have a polytunnel on a sloping site, but great care needs to be taken with its installation, especially with the orientation of the hoops. Your supplier should be able to advise.

Exposure Avoid an exposed, windy spot. If possible, choose a sunny site, far enough from any trees to eliminate any shading.

Size It is generally worth going as large as your site and budget allow, but be aware that you may require planning permission, especially if the polytunnel will be close to houses or a road, and it may need some form of screening. Taller tunnels are more useful, as they have plenty of head room and space to hang things, but if your site is overlooked or is visible from a road, you may have to choose a lower one. Don't forget that large polytunnels create a lot of water run-off and, in winter, a lot of snow may have to be cleared to avoid damage. You can buy second-hand hoops to make cost savings. When buying your frame, buy bracing bars that allow you to hang ropes and suspend pots, and which give more support in exposed windy locations.

Style The classic shape is a semi-circular hoop, but you can get tunnels with straight sides, providing more workable space, or ones with a more greenhouse-like profile.

Covering There are now several different kinds of covering available, most of which are polythene. Replacing the cover is a big job, so don't economize on the polythene: a good-quality UV-stable cover can last 5 years or more. Avoid the temptation to buy cheap polythene from a builder's merchant, as it won't be UV-stabilized and will only last for a couple of years. Ideally, the thickness of the polythene should be 200 micron (800 gauge). For a bit extra, you can get an anti-fog polythene that reduces condensation and has improved heat retention. White polythene provides a cooler environment, which may suit some crops. Also, it is useful for tunnels where animals

It's worth getting as large a polytunnel as your site and budget will allow, but be aware that you may require planning permission.

may be housed in winter, as there is less risk of the tunnel getting too hot on sunny days.

If you have an exposed site, where a traditional polytunnel may get damaged, you might choose laminated plastic – which is a bit like bubble wrap. The laminations trap air, providing improved insulation and considerable strength: these covers can survive gales and heavy snow loads, and you can even walk on them! This plastic and framework may cost more, but the sections can be replaced and extensions added. The high heat retention extends the growing season without the need for supplementary heating, and the plastic also provides more diffuse lighting, as the light is scattered as it passes through the bubbles.

Cover fixing The cover has to be secured in place, and this is most commonly achieved by erecting the frame and then simply burying the plastic in a trench. Alternatively, the plastic can be attached to a wooden base rail. This is an easier option and makes for rapid replacement of the plastic when needed. However, the rails don't prevent small furry mammals from digging underneath!

Doors You will need large doors at each end that can be opened for access and ventilation. The latter is particularly important for a longer tunnel, where a lack of air movement can cause problems and lead to the build-up of pests and disease.

Ventilation Temperatures in summer can exceed 40°C (104°F), even with the doors open. This can be reduced by shading the structure or by spraying water inside, though this is not ideal as some plants don't like a humid environment. One alternative is a polytunnel with sides that can be raised in summer and the opening covered with insect mesh. We have a second set of doors on each end of our tunnel, made from insect mesh, for the same purpose.

Access The polytunnel will be visited daily, so it needs to be easily accessible and, if possible, not at the far end of the plot, as you may want to bring in water and electrics, as well as barrowing in loads of compost and carrying away your harvest. There should be a clear area around it, ideally under a ground-cover fabric to stop weeds growing, because you will have to replace the plastic at some point and, if you have beds right up to the plastic, you won't be able to do this easily. You will also need access for minor repairs and for washing down the cover.

Accessories You'll need some super-thick sticky repair tape, and some anti-hotspot foam tape to cover the frame, so there is no direct contact between the plastic and the metal.

If it's not possible to fence your whole plot with rabbit netting, at least use it to fence your growing area.

Solar tunnels

An alternative to an ordinary polytunnel is a solar tunnel. This is a hybrid polytunnel-cum-greenhouse, and it is particularly useful on sites where a large area of plastic would be seen to be unattractive. Solar tunnels are stronger than ordinary plastic ones, as they are formed from a double layer of plastic with mesh reinforcement between. They have base rails and are often modular, so they can be extended and are easier to relocate. The investment is greater but you end up with a more flexible tunnel.

Controlling pests

As soon as you start to keep animals, especially poultry, the pests arrive. Rats are the main problem for many smallholders, attracted by the free source of food. Always keep feed in a rat-proof feed store or use metal bins. The use of automatic feed dispensers will help to discourage them too, as food is not on the ground. You will never to be able to get rid of rats completely, but you do have to keep them under control. Rats can be poisoned, shot or trapped. You will probably use all these methods at some point! Poison needs to be put down in bait boxes in places where livestock cannot reach it, and any dead rats must be disposed of. An air rifle can be a good way of killing the rats that emerge to feed on poultry food, and often you can find a local person happy to come and shoot for free. In areas where you do not want to put down poison you can use traps. You can buy snap-shut-style fen traps, to place in roof spaces and where there are no other animals to get trapped, or live traps.

Foxes cause problems for poultry keepers. To be fox-proof, fencing needs to be either electric or at least 1.8m (6') high with a reinforced base or outward-facing chicken wire buried in the ground to stop the fox burrowing underneath – or a combination of both. Live traps can be used too, and foxes can be shot at night.

The abundance of tasty vegetables will draw in rabbits. If you can afford it at the start, include netting along your perimeter fence to keep them out, and put wires across the bottom of your gates. Rabbit netting has a small mesh (2.5cm/1") and is at least 60cm (2') high. It is supported with posts or stakes, and the lower

The vegetable-growing area on this holding is protected by electric deer-proof fencing, comprising lines of polytape.

15cm (6") is buried or bent outwards to stop rabbits from burrowing underneath. If it's not possible to fence your whole plot with rabbit netting, fence just your growing area. You are vulnerable without netting, but some people swear by scattering blood and bone around crops. Not only is this a good source of long-term nutrients, such as phosphorus and potassium, but also the smell deters the rabbits, as does wood mulch around valuable crops.

Other pests include corvids (jackdaws, magpies and crows), which are attracted to livestock feed. They can be shot or caught in live traps such as the Larsen trap. Buzzards, red kites and other birds of prey can take young birds, but in the UK they are protected species. They cannot be shot, so the best way to protect young birds is through the use of nets over pens. The nets will discourage corvids too.

Deer can be problematic where there are orchards and other young trees, and the only way to keep them out is to invest in deer-proof fencing around the perimeter: either 2m (6'6")-high netting or an electric fencing system to a similar height, with visible polytapes.

chapter TWO

Soil matters

Soil is a living entity, and is the most important asset of any plot. On a small plot, the soil is going to be worked particularly hard, so it's essential to understand how it functions, how best to cultivate it, and what you can do to improve it.

A section through a loamy clay soil, with dark brown topsoil over a lighter brown subsoil and a thick band of clay beneath.

Teeming with life

Just one teaspoonful of soil can contain several hundred nematode worms, a million single-celled organisms called protists, a million individual fungi and a billion bacterial cells of thousands of different species.

A typical definition of soil is that it is the top layer of the earth, consisting of rock particles, organic matter, air and water – but soil is much more than that. It's a living entity that teems with life, from worms and beetles to microscopic bacteria and fungi. For soil to be healthy, it needs plenty of air and water for all these organisms, and nutrients for plant growth. These nutrients come from the minerals in the soil and from the breakdown of dead and decaying matter in the soil: a process made possible by decomposers such as bacteria and fungi. The more nutrients in the soil, the more plant growth, which in turn means more food for animals and, eventually, more food for us.

The way you manage the soil on your plot affects soil life, so you need to think carefully about how you are going to cultivate your vegetable area. Will you go down the traditional 'digging' route or opt for a no-dig approach?

Soil basics

The particles that make up soil – sand, silt and clay – are grouped according to their size. Sand particles are the largest and clay particles the smallest. Most soils are a combination of the three, and it is the proportion of each that gives soil its texture. A soil that is mostly sand particles has lots of air spaces, and this creates a light, free-draining soil which is prone to drought. One that is mostly clay particles is much heavier, with fewer air spaces, and has poorer drainage. It is slow to warm up in spring and bakes hard in summer. Loamy soils are the most desirable, with roughly equal parts of sand, silt and clay.

Colour reveals a lot of information about a soil too. Soils that are rich in organic matter are dark brown or even black. A soil that is rich in iron is orange-red in colour, while a soil that is frequently waterlogged has a mottled yellow-grey appearance.

Soils are rich in mineral nutrients. The three most important of these are nitrogen (N), phosphorus (P) and potassium (K). They are described as macronutrients, because they are required in larger quantities than other nutrients. However, if any of the other nutrients, such as calcium, sulphur or magnesium, are in short supply, plant growth will be restricted.

Cotswold brash is a clay loam with a high stone content. The reddish-brown colour comes from its mineral content.

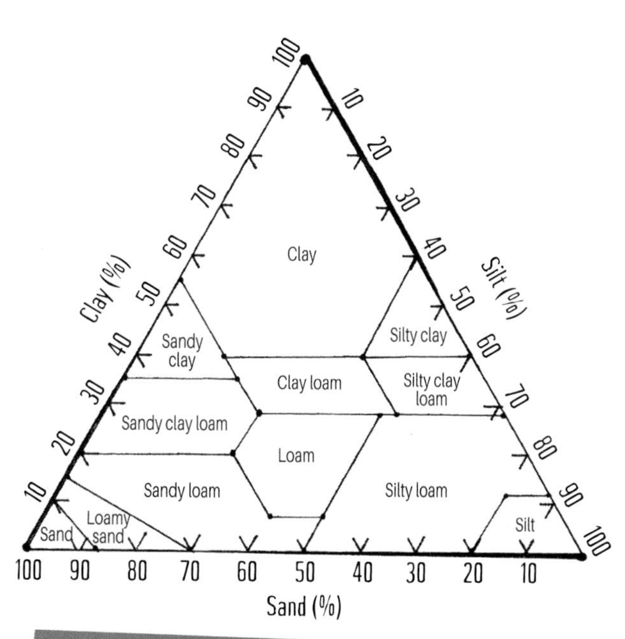

This diagram shows how the mix of sand, silt and clay varies in the different soil types.

The so-called 'micronutrients', such as iron, manganese, zinc, molybdenum and boron, are needed in very small quantities.

Soil pH

'Soil pH' refers to the acidity or alkalinity of the soil. Soils can be naturally acidic or alkaline, according to the proportion of minerals in the soil. An alkaline soil has high levels of calcium and magnesium, for example in the case of soils lying over chalk, while sandy soils tend to be acidic and have more iron present. These soils do support plant growth, but they are not ideal for cropping. The best soils are loams rich in organic matter, with a neutral pH of around 6.5 to 7 (the higher the pH value, the more alkaline the soil).

Potatoes like a rich, well-drained soil. A slightly acidic soil helps to reduce the risk of scab.

Soil pH is important, as it will affect the availability of minerals. In the neutral pH range, all of the soil minerals are available to the plants, but if the pH is increased or decreased their availability can be reduced. The pH of a soil can alter over time: for example, if crops are continually removed and the nutrients are not replaced, or if there is leaching of nitrates (a soluble form of nitrogen). The continual use of artificial fertilizers will also alter soil pH. Soil life is affected by pH too. Acidic soils tend to support fewer fungi and bacteria, and important groups of animals such as earthworms and nematodes disappear.

Organic matter

'Soil organic matter' refers to the remains of living organisms that are incorporated into the soil, releasing nutrients as they decompose. For hundreds of years, farmers have added farmyard manure and compost to their soils to replace the nutrients taken up by their crops. The bulky, fibrous nature of organic matter helps to improve the soil structure, reduces compaction, and improves aeration and water drainage. Organic matter also acts as a carbon store. So, it's critical on your plot that you continue to add organic matter to your soil, otherwise you will have problems maintaining productivity. In sandy soils, organic matter improves soil's ability to retain water, while in heavy clay soils it will improve drainage. Some organic matter does not break down and instead forms part of the soil, where it holds on to nutrients and prevents them from leaching away. For more on the topic of organic matter and its role in composting and soil fertility, see Chapter 3 and further information in Resources.

Soil life

It is estimated that as much as a quarter of all life on land may exist in the soil. Dominated by bacteria and fungi, there are also protists, nematodes, earthworms, insects, spiders, mites and much more. Soil organisms have many roles: decomposition of organic matter, nutrient cycling, maintenance and improvement of soil structure, soil drainage and water retention. This amazing web of life is critical to the health of soil, but surprisingly little is known about it or about how it is affected by practices such as adding inorganic fertilizers and using pesticides such as fungicides.

When you have lots of microorganisms in soil, you tend to find that the 'good' ones keep the 'bad' ones at bay.

Soil organisms are critical to the process of soil formation. As organic matter breaks down, it releases complex polysaccharides and gums, which stick tiny rock particles together to form micro-aggregates, and these in turn are then bound together by plant roots and fungal hyphae to form larger particles. Large animals, such as worms and insects, move these particles around, creating and recreating pores in the soil, which improves aeration and drainage.

A healthy soil with high organic matter content usually has a large and diverse population of microorganisms. When you have lots of microorganisms in soil, you tend to find that the 'good' microorganisms keep the 'bad' ones at bay, so you have less disease and fewer pests. Sir Albert Howard, one of the pioneers of the organic movement, understood the link between healthy soil, healthy food and healthy people. His years of research into soil and composting led him to conclude that the lower the state of fertility of the soil, the greater the likelihood of pests and diseases.

Bacteria and other microorganisms release enzymes that break down and change nutrients in soil particles into a form that plants can take up. In fact, these microorganisms are so important that the plants feed them. Up to half of a plant's photosynthetic product is moved to the roots, where it is exuded into the soil to feed the organisms that live around its roots in the zone called the rhizosphere.

There is a whole food web in the soil around plant roots. The bacteria are eaten by protozoans (single-celled organisms such as the amoeba), which are eaten by nematodes and micro-arthropods, and these in turn are eaten by still larger organisms. All these organisms excrete waste materials into the soil, which are taken up by the plant roots, and when they die they decompose and nutrients are released.

Earthworms

Earthworms are critical to soil health, as their burrowing habit breaks up the soil and helps water to drain through the soil, and allows oxygen to reach plant roots. Earthworms pull

Earthworms burrow through the soil, mixing organic matter and improving aeration and drainage.

dead leaves from the surface into their burrows, where the leaves are eaten and digested. As a result, a soil with a healthy population of worms has a better soil structure and more nutrients, which leads to improved plant growth.

However, worms are sensitive to compaction. Walking on the soil or driving over it with heavy vehicles presses the particles together, squeezing the air spaces and leading to poorer drainage. This makes it more difficult for the worms to move through the soil. Even the action of digging the soil destroys their burrows and causes them disturbance (not to mention chopping up the worms). Sometimes, of course, disturbing the soil is unavoidable, but it is important to minimize disturbance, and this is the value of the no-dig approach (see page 41).

Clearing and cultivating

If you have taken over a derelict corner of a field or piece of overgrown land, the prospect of trying to grow vegetables on it can seem quite daunting. So your first job is to halt the weed growth and get it back into some order. Then you will need to prepare the ground for sowing or planting.

Before you start any work, walk the site and make a note of the weeds that are growing. Tall and lush weeds, such as nettles, indicate good soil fertility. Buttercups, daisies and docks tend to prefer slightly acidic soil. Docks and horsetails like it damp. Take a spade with you so you

On this allotment, some of the beds have been prepared, but the others are left under covers to stop weeds germinating.

How to check soil structure

Have you ever had a good look at the soils on your plot? By digging a hole in the soil and taking out a wedge you can check the structure and look for any problems. Soils with good structure will have well-formed porous blocks with rounded edges, which are easily broken with the fingers when moist. Vertical gaps through the blocks allow the roots to grow deep into the soil. Drop the wedge of soil to the ground to see how easily it breaks up. Soils with poor structure have harder and sharper blocks which are difficult to break apart. These soils are easy to damage.

A clay loam with plant roots extending deep into the soil.

can dig a pit and have a good look at the soil. Then clear away any rubbish, cut down any unwanted woody plants, and cover as much as possible with a light-excluding weed barrier such as tarpaulin, permeable ground-cover fabric or even a large piece of plastic (agricultural silage plastic is the cheapest to buy), which will stop any new weeds growing and, in time, kill the existing weeds. This type of weed barrier may need to be left in place for as long as a year or possibly more, depending on the type of weeds, but at least you have the process of control under way. On my own plot, the bindweed survived a year under silage plastic, so I had to leave it down for a further year. The weed barrier is a stop-gap until you have time to cultivate your beds. If you cannot leave the weed barrier in place for long enough for it to do its work, then uncover bits at a time, leaving the rest in place.

Digging

The classic scene of vegetable garden on a sunny day in winter is one of a gardener digging his or her soil. It's hard, back-breaking work, especially on clay soils. Cultivating through digging, either manually or with a rotovator, may be a common approach, but it's not always the best one and often it's not needed. Every time the soil is turned, soil organisms are disturbed or harmed.

Digging a bed of a heavy clay soil is hard work.

Another problem is that after the soil is tilled, nutrients are released and organic matter breaks down, which means there is less for your crops. Also, turning the soil brings seeds to the surface, so you get a flush of weeds, while beneficial organisms in the soil such as worms and fungi are damaged or destroyed.

Soils are usually dug over when beds are being prepared from scratch or at the end of the growing season to clear crops. The process of manual digging involves taking a spadeful of soil at a time and turning it over to bury weeds, so they rot in the soil and add to the organic matter. This is followed by raking to break down the clods of soil and make a fine tilth that can be sown or planted into.

For a new bed, especially one on a compacted soil or heavy clay soil, the traditional method is to double-dig. This improves drainage and aeration, and loosens a sizeable depth of soil for vegetable growing. Double-digging involves digging a trench to the depth of a spade at one end of a marked-out bed, putting the soil into a wheelbarrow. Then the soil at the bottom of the trench is forked over (hence 'double-dig'), and at this point organic matter such as compost can be incorporated. Then a second trench is dug beside the first, with the soil piled into the first trench, and the process of forking over the bottom is repeated. This process is repeated along the bed. When you get to the end of the bed, the final trench is backfilled with soil from the wheelbarrow.

Rotovating

Another popular method is to use a rotovator, but there are drawbacks with that approach too! It's a speedy method that prepares the ground with minimum effort, turning a virgin soil into a fine tilth ready to be planted. However, the blades will chop up the roots of perennial weeds, such as docks and bindweed, which then reappear in vast numbers a few weeks later. A rotovator tends not to get down more than 15cm (6"), and if used on very wet soils or on a clay soil it can create a hard pan just below the level of the blades. This pan is a compacted layer of soil which prevents the downward movement of water, leading to waterlogging and making it difficult for roots to penetrate.

Using chemicals

One quick way to get rid of weeds is to use weedkillers such as glyphosate, which will kill all but the most stubborn weeds in a few weeks. Although glyphosate may not leave a residue in the ground and the soil can be worked straight away, there is mounting evidence that it kills soil life and is also only a short-term fix: often, weeds reappear a few weeks later.

These beds of onions and chard have been no-dig for many years.

The no-dig method mirrors what goes on in natural systems, where the soil organisms are undisturbed.

The no-dig approach

The alternative to the methods just described is a no-dig or no-till system, in which compost is spread on to the surface of the soil, to be gradually incorporated by the action of worms and other soil organisms. This mirrors what goes on in natural systems such as woodlands, and, with no turning of the soil, the soil organisms are undisturbed. The compost provides them with a supply of nutrients and, as decomposition takes place, the soil life increases, while larger organisms such as earthworms help to aerate the soil and improve drainage.

Getting started with no-dig on an overgrown piece of land is probably the hardest and the most time-consuming part. It involves putting down a thick layer of mulch (top dressing) to smother the vegetation. First, mark out your beds,

Once the crops have been cleared, we spread a thin layer of compost over our no-dig beds.

but don't remove the weeds, as you must keep the soil intact. The mulch will need to be thick, so it can help to edge the beds with boards to keep it in place. Strim or chop off all the vegetation as low as you can get it, and leave the clippings on the surface. Next comes your weed barrier, which can be a layer of thick cardboard or newspaper. I prefer cardboard, as I have found that newspaper rots down too quickly on beds with pernicious weeds such as nettles. Give this a good soaking. This barrier will rot in time, but not before the plants underneath have died.

Now you can start building up the layers of mulch. It's a bit like building a compost heap (see Chapter 3). Each layer needs to be about 5cm (2") in depth, with roughly one layer of 'green' materials (lawn clippings, green garden waste and so on) to three layers of 'brown' (woody materials, cardboard, newspaper, etc. – for more details see page 53). If you want to kick-start the process, start with more 'green'. As you are building the layers, add an occasional sprinkle of lime to keep the pH fairly neutral. The final depth of mulch should be around

Pros and cons of no-dig

The no-dig method has many advantages:

It protects the soil structure.

It creates a soil with a rich, friable surface, which is less likely to cap and so force water to run off.

Moisture loss is reduced.

Weed seeds are not brought to the surface to germinate.

It helps to create a rich soil life, especially in the case of worms, which are not disturbed by digging.

It allows the soil to be worked on all year round.

There is no need for back-breaking digging.

But also some disadvantages:

Soil pests are not exposed to predators and cold weather.

It doesn't deal with compaction or hard pans.

It takes longer to clear the ground of weeds.

The soil can take longer to warm up in spring.

Some people miss the therapeutic benefits and exercise associated with digging.

Creating a no-dig bed on overgrown land involves adding layers of mulch. It's a bit like building a compost heap.

30cm (12"), but if you want deeper beds, you will just need more mulching materials! As the materials decompose, the mulch shrinks by a third or even more. Ideally, you need to leave the beds to settle and for the decomposition to get under way before using them, but some people use them straight away. If you do this, remember that the decomposition will generate some heat, so it's best to use transplants rather than sowing seeds direct. Dig a small hole for the plant and backfill it with compost. A newly mulched bed is really good for squash and courgettes, as there is plenty of moisture and ample nutrients to boost their growth.

To convert existing vegetable beds to no-dig, all you need to do is add a layer of compost in autumn. Come spring, the compost will have been mixed in by the action of worms and other soil life. Any weeds that appear can be pulled or hoed off. You'll need to hoe regularly to kill weed seedlings, or, if the weeds have got away, smother them with some mulch. The only time you need to move any soil is to make a hole for planting.

Chapter THREE

Soil fertility & crop rotations

Over time, the removal of crops from the land slowly depletes the soil minerals, especially phosphorus, so you need to add nutrients in order to keep your soil healthy. Adding nutrients in the form of organic matter also helps to support more soil organisms, which are important for making nutrients available to plants.

It is particularly important to look after the soil on your heavily cropped vegetable beds and in the polytunnel or greenhouse, where year-round cropping can take a heavy toll. Organic matter can be added as compost or manure, as well as by growing green manures. Compost in particular is key to a healthy soil, and this chapter describes the composting process and methods in detail. It also discusses crop rotations: the system of growing crops in different beds from one year to another, rather than in the same bed year after year. This traditional approach has long been used to prevent the build-up of disease and pests.

Feeding the soil

It's very easy and tempting just to add artificial fertilizers to your soil to provide nutrients. Artificial fertilizers contain nutrients in a form that is readily available to plants, for example as nitrates, so when they are added to the soil they are taken straight up by the plant roots, bypassing the soil completely. It's no surprise that over time, soil organisms and soil organic matter decrease and the soil becomes simply a medium to anchor the plant roots. As a result, it becomes necessary to add artificial fertilizer every year.

A much better way is to feed the soil itself, by adding organic matter. Growing green manures – plants grown as a ground cover and later

All crops take up nutrients from the soil as they grow, so those nutrients must be replaced.

It's very easy and tempting just to add artificial fertilizers to your soil to provide nutrients, but over time the soil life and soil organic matter will decrease.

incorporated into the soil – is also beneficial. These all provide food for soil organisms, especially the decomposers such as bacteria and fungi. There are other benefits too: a soil rich in organic matter will retain more water and nutrients and have a rich and diverse soil food web, making nutrients available to plants and helping to suppress disease.

Using composts and manures

You can use a range of organic materials to boost fertility, such as homemade garden compost and various farmyard manures. Composts and manures vary in their pH and their nutrient content.

Fresh animal manure has much of its NPK (nitrogen, phosphorus, potassium) content in readily available form, so great care should be taken if it is used on the plot. The abundance of nutrients in a fresh manure can cause more harm than good, as the high nutrient levels can depress the activity of microorganisms, and the soluble nitrogen is readily leached from the soil and may even pollute nearby watercourses. It is far better to use a well-rotted manure, which will help to build up soil fertility over a long period of time, but will not feed the plants directly. The description 'well-rotted' indicates that it has been allowed to age and break down, and can be safely added to soil.

Composts are made from a mix of organic materials. They have been stabilized (broken down into new compounds that may be harder to decompose, and some of these become bound to the soil particles), so there is more dry matter. The percentages of NPK in readily available form are much lower, so there is much less risk of nitrogen leaching. Composts tend to be a good source of potassium and phosphorus, providing a longer-term supply of nutrients. They are bulky, so are usually added as a mulch, or dug in as a long-term soil improver.

This new raised bed has been filled with well-rotted horse manure.

The end product of decomposition is humus: a brown, crumbly, organic material that is rich in nutrients.

The composting process

Composting is the breakdown of organic materials by microorganisms, and is an essential part of the biological cycle of growth and decay. It is readily visible in natural habitats – look at a woodland floor and you will see a layer of leaf litter. It doesn't stay there long, because the bacteria, fungi and other microorganisms break down the complex organic compounds such as proteins into simpler inorganic ones such as nitrates and phosphates, which plants can use. The end product of decomposition is humus: a brown, crumbly, organic material rich in nutrients and soil life.

There are three phases to composting. The first stage is when the materials start to break down as a result of the activity of the microorganisms, especially bacteria. This process generates heat. The temperature rises steadily and after a few days will exceed 50°C (122°F). This temperature kills off the first-stage microorganisms, and they are replaced by others that can tolerate higher temperatures. To kill off pathogen organisms in the compost you need to achieve at least 55°C (131°F), but if it rises above 65°C (149°F), beneficial organisms will be killed too, so the compost has to be turned and aerated, lowering the temperature. This mixing also helps to incorporate more oxygen, keeping all the breakdown processes aerobic. This high-temperature stage can last for days, weeks and in some cases months. The final stage is when the activity of the microorganisms declines as the supply of high-energy compounds has been exhausted. The temperature falls and the first set of microorganisms takes over again, and other organisms arrive in the mix, such as woodlice and tiger worms.

The microorganisms involved in composting include bacteria of many different types, each adapted to working at a different temperature. They release enzymes that digest many substances, in particular the nitrogen-rich foods. The fungi and actinomycetes (a type of bacteria) are capable of breaking down tough woody materials. The actinomycetes are related to fungi, and they give soil its characteristic earthy smell. Fungi and actinomycetes are particularly important in the final stages of composting, when all that is left is the tough woody material rich in lignin and cellulose, and they help to form the humus.

No smells!

It is important during composting to make sure there is plenty of oxygen for the beneficial microorganisms to function. As soon as the oxygen levels fall or there are pockets that lack oxygen, the process of decomposition slows down. When there are anaerobic (oxygen-free) conditions, other microorganisms take over. They work more slowly than the aerobic ones, do not generate much heat, and their by-products are smelly gases. If your compost heap is turning into a putrid, smelly mess, you have anaerobic microorganisms at work.

These fallen leaves will soon be decomposed by bacteria and fungi.

Dead leaves are high in carbon and, in the absence of other materials, break down very slowly. The resulting leafmould is low in nutrients but is a valuable soil improver.

Carbon and nitrogen

When people talk about compost, they often discuss the importance of the carbon-to-nitrogen ratio (C:N) and how it affects the breakdown process. Carbon makes up much of a plant's support structures – the cellulose in the cell walls and the lignin in the woody parts – while nitrogen is an essential component of proteins, which are found in all cells. Cardboard, paper and woodchip all have a high C:N, of around 400:1, because they are made from woody fibres, while animal manure, grass clippings and other lush green plant material is rich in nitrogen, with a C:N of 25:1 or less. The optimum C:N for creating a sweet-smelling, fertile compost is 25-30:1, but this ratio doesn't exist in many plant materials, so they have to be mixed – 'green' (high in nitrogen) with 'brown' (high in carbon). If there is too much carbon, decomposition will be slow, but if there is too much nitrogen, as in a pile of grass clippings, decomposition will be rapid and you get left with a smelly, slimy mush.

The balance of C:N is critical to soil life too. Compost with a lot of visible (undecomposed) woody material has a high carbon content, and when you spread this compost on soil it can create problems. To break down the carbon, the microorganisms use the nitrogen in the soil, depriving the plants (this is known as 'nitrogen robbery'). To stop this happening, you need to make sure that the materials you add have a good balance of carbon to nitrogen – a C:N of about 25-30:1. If you have compost with a lot of undecomposed woody pieces, you should add it

compost colour

Colour tells you a lot about the quality of a compost. The final compost should be dark brown: aim for the colour of a 70%-cocoa chocolate! If it gets too hot, the compost will be very dark brown, almost black, indicating that anaerobic conditions were experienced, so the compost will lack nutrients and microorganisms. A white, ashy layer indicates that there was compaction in the compost heap, which can be caused by having too much fine material.

The composting process can be completed in 6 months or less by using the right mix of materials and turning the heap several times.

to the soil in autumn, to give the microorganisms all winter to finish the decomposition and to allow time for the nitrogen levels in the soil to build up again.

Making compost

Composting is an absolutely essential element of a productive small plot, allowing you to recycle all the organic matter, so you can return nutrients to the soil for the next crop of plants. Whole books have been written on the composting process and everybody has their preferred way, so here I will just touch on the basics of making a good compost.

Composting takes time. Many of us, I suspect, have a compost heap in the corner of the vegetable garden where we toss the weeds and grass clippings and let nature take its course. That will work . . . it's just that it will take time and often creates a compost that is full of weed seeds and pieces of twigs and tougher materials. Alternatively, the process can be completed in 6 months or less by using the right mix of materials and turning it several times.

On my plot I have a number of compost heaps. In my raised-bed area, I simply designate a bed as a compost heap and pile up the materials as I weed, trying to have a mix of about a third to a half 'green' and up to two-thirds 'brown', adding shredded paper, cardboard or straw as needed to balance the ratio. The green material will break down quickly, uses up oxygen and generates some heat, which will fuel the decomposition

A classic row of compost bins in Perth, Australia.

A compost bin made from pallets has plenty of gaps to aerate the heap as it builds up. The slats across the front have been removed while the bin is filled.

process. I cover it up with black plastic and each week add fresh material from my weeding and pruning. By the end of the growing season I have a good-sized heap, which I leave over winter and use the next year for the hungry crops such as potatoes. In this system, the temperatures may not rise high enough to kill the weed seeds, so there will be a good flush of weeds, which I hoe off as they germinate.

In addition, I have a traditional row of compost bins made from recycled pallets, which I fill in much the same way as on the raised bed. The slats allow air into the compost – an essential element for composting, as it keeps the process aerobic, and I don't keep them covered all the time, so water can reach the heap too (it shouldn't be too dry but not too wet either!).

By carefully controlling the mix of composting materials, you can produce a compost that is either rich in bacteria or rich in fungi, which can be put to different uses. A bacteria-rich compost is one that has had more nitrogen-rich materials present, such as legumes, grass clippings and manure. These are rich in protein and sugars, which fuel the growth of bacteria. The resulting compost is good for mulching soils to grow vegetables (especially brassicas), flowers and herbs. It's also the preferred compost for making compost teas (see page 55).

A compost that is rich in woody material has a high fungal component, and is good for mulching fruit trees and bushes, new hedgerows and other perennial plants.

Materials that can be used for composting

'Brown' materials (high in carbon – C:N 50-600:1)	
Bracken (C:N of 50:1)	Can be readily obtained in some areas.
Cardboard, newspaper and shredded paper (C:N of 350+:1)	These should be torn or shredded to create a larger surface area for the microbes to work on, and dampened. Avoid finished glossy papers, which don't rot down easily. Sheets of cardboard and newspaper can also be used as a weed barrier at the bottom of a new no-dig bed, dampened and covered with other mulching materials.
Dead leaves (C:N of 40-80:1)	Readily available from many gardens.
Straw (C:N of 50-150:1)	Straw is dried stems, with little sugar content. You may be able to get spoiled bales of straw for free locally.
Wood shavings, woodchip, sawdust (highest C:N, of 300-600:1)	Your local timber yard or tree surgeon may be able to provide these for free, but avoid treated wood and yew. Wood chippings are also great for mulches and on hugelkultur beds (see page 54).
Woody garden waste (C:N of 50-400:1)	Shred your woody garden waste, as smaller pieces will break down more quickly.

'Green' materials (high in nitrogen – C:N 5-25:1)	
Animal manure (C:N of 5-20:1)	Animal manure has a high nitrogen content. You can use poultry manure from your chicken houses or buy manure in from local farms or stables.
Coffee grounds and teabags (C:N of 20:1)	Easily obtained and often available from coffee shops.
Comfrey leaves (C:N of 10:1)	The leaves of this excellent plant can be cut and used as a mulch, added to the compost heap or rotted in water to make a foliar feed. See page 57 for more details.
Grass clippings (C:N of 9-25:1)	A readily available green material. Be careful not to add too thick a layer of clippings, as they can rot down to create a dense, impenetrable mat. Ideally, mix with shredded paper or newspaper, or cardboard.
Green garden waste (C:N of 5-25:1)	Weeds and prunings from the garden.
Hay (C:N of 15-25:1)	Hay is rich in sugar, which is good for feeding bacteria in the heap. Dampen to activate this process. On the downside, hay is likely to contain seeds, which will germinate, and if used as a mulch it can encourage slugs and snails.
Seaweed (C:N of 5-25:1)	This is high in nutrients and makes an excellent mulch as well as compost activator. If you live near or have easy access to the coast, you may be able to harvest seaweed from the shore for free, especially after a storm, but check that there are no local laws restricting its collection.
Other materials	Other 'green' materials include old cut flowers, vegetable and fruit waste, citrus fruit skins, pondweed, spent hops, wool (fleece from shearing), shredded silk clothing – and urine!

Do *not* add the following materials to your compost heap:
Diseased plants; persistent weeds (such as bindweed and ground elder, which regrow from tiny pieces); soot and coal ash; dog or cat faeces and cat litter; glossy paper; meat, fish, fat and cooked foods (which smell and attract vermin); grass clippings from lawns treated with weedkiller.

Hugelkultur beds

Hugelkultur (hoo-gul-culture) beds have been used for hundreds of years in Eastern Europe. They are heaped or mound beds – neither raised beds nor compost heaps, but something in between – and make use of woody waste, that is fallen branches and logs. They are great for long-term fertility, moisture retention and increasing the surface area available for planting. As they contain a lot of rotting wood, there is a supply of nutrients for crop plants for 10 to 20 years, plus the decomposition generates heat, which gives the beds an early start in spring. There is excellent aeration from the plentiful air spaces between the logs and branches, while water is retained rather like in a sponge. This means that there is little need to water the bed in all but the driest summers.

These beds are built by first digging out a trench about 30cm (12") deep, which is filled with logs and thick branches. These are covered by smaller branches and woodchip, then topped with compost or well-rotted manure. Some people prefer to make steep sides to the beds so that compaction is avoided; others prefer a more shallow bed. It's all a matter of preference. Since the logs and branches are carbon-rich materials, they take up nitrogen as they decompose, so it's important to have a thick layer of nitrogen-rich compost and initially to grow nitrogen-fixing plants such as legumes. Once the wood starts to decay, nitrogen is released. To kick-start the decomposition process, some people add green matter, leafmould or seaweed around the logs.

Our small hugelkultur bed has a base of logs and branches, which were then covered with woodchip and a thick layer of compost.

Sometimes, your compost heap needs a little help. You may have a lot of brown material and not enough green, or the weather may be cool and the microbiological activity low. In these circumstances you can use activators – substances that are high in nitrogen – to kick-start the whole process. Activators include bloodmeal, bonemeal, comfrey tea (see page 57), fresh animal manure, seaweed and urine.

Compost teas

A compost 'tea' is another source of nutrients and can be used as a boost for cropping plants. It also helps to beat disease-causing organisms by adding beneficial ones. It is made by adding some good bacteria-rich compost or some vermicompost (worm compost – see below) to a container of water for a couple of days, and the resultant liquid is used as a foliar spray. Some users claim that a compost tea can boost yield by 10-20 per cent.

To make a tea you will need a large container, such as a 20-litre bucket (approx. 4 gallons / 5 US gallons), an aquarium pump, tubing, air stones, a long stick for stirring, molasses, a source of microorganisms such as a good compost, and something to strain the tea, such as old tights or a pillowcase. Aeration is key, as the organisms will quickly use up the oxygen and the liquid will become anaerobic. If you are using tap water, let it stand for a day or so to get rid of the chlorine. Half-fill your container with your compost and connect the air stones to the tubing and pump, making sure they extend to the bottom of the bucket. Then, add your water so that it comes to 10cm (4") below the rim of the container. Start the pump and aerate the liquid. After 10 minutes, add about 2 tablespoons of molasses and stir. This will feed the bacteria and boost their growth. Leave to bubble for 2 to 3 days, stirring vigorously a couple of times each day to stir up the compost. Don't leave for any longer without feeding again. Turn off the pump, allow the compost to settle and strain the liquid to another container. You must use the tea straight away, while it is full of beneficial microorganisms, and don't waste the solids. They can be mixed into soil or popped back on to the compost heap.

A bacteria-rich compost tea is claimed to be effective as a foliar spray to combat many diseases, especially those carried by airborne spores that land on leaves, such as mildews and blight. But don't spray compost tea directly on to leaf crops that are to be harvested within 2 to 3 weeks, as there could potentially be harmful bacteria in the liquid. Compost tea can be applied to seedlings at the two-leaf stage to reduce disease such as damping off, and then again at 2 months to give them protection from other diseases and pests.

Wormeries

Kitchen waste cannot be fed to your animals, because of the risk of transmitting disease, but much of it can be either popped on your compost heap or fed to worms. A wormery is simply the name for a colony of worms living in a container and feeding on organic matter. Their activity converts the kitchen waste – such as

This wormery consists of several layers. The kitchen waste is added to the top layer, where the tiger worms are to be found. When a layer is full, a new one is added on top.

Tiger worms, also known as brandling worms.

vegetable peelings and rotten fruit, bread and cooked food such as pasta – into a fine compost, while the water content of the fruit and vegetable matter is released and can be collected as a nutrient-rich liquid to be used as a plant feed.

These worms can be kept in any one of a range of different types of container, from an old plastic dustbin or water butt to a purpose-made multilayered system with a sump for collecting the liquid waste. A wormery needs to be kept in a sheltered place, out of direct sunlight, where the temperatures will be constant: for example, a garage, a shed, under a lean-to or in a shady spot on the plot.

Compost worms are different from earthworms, which prefer soil. Of the several types of worm found in compost, those used in wormeries are usually tiger worms, also known as brandlings (*Eisenia fetida*). When setting up a wormery from new, you can order a supply of worms to get the whole process started. If you have a multilayered system, you place some old compost or coir as bedding on the bottom layer and add the worms. Water it a little, so it is moist but not wet. Then add the kitchen waste, starting with a thin layer. Then, at weekly intervals, add more food, small amounts at a time and preferably shredded, as this is easier for the worms to manage. Don't keep on adding food if the worms are not coping with it – a thick layer of rotting food will only create a stench and attract flies.

Composting worms tackle most foods, but you should avoid citrus peel, spicy foods, onions, milk products, fats and oils, and any meat or fish, as well as tough materials like nuts or woody stems, as these will be avoided by the worms. To prevent too much liquid coming off, add about 20-30 per cent brown waste such as shredded newspaper, cardboard or even wood shavings. Once one layer is full, add another layer on top. The worms will move up as they search for a fresh supply of food, leaving a fine compost behind. A liquid will start to collect after about 10 to 12 weeks, which needs to be drained off regularly. You can use this as a liquid feed, diluted with water to a ratio of 1:10, as it is very concentrated. The vermicompost that the worms produce is really good for making compost teas (see page 55).

Comfrey

Comfrey is a must on any plot. This attractive plant, a relative of borage, is known as a 'bio-accumulator' because its roots grow deep into the subsoil and draw up nutrients. The result is leaves with a particularly high nutrient content, especially NPK. Analysis shows the leaves to be about 17 per cent nitrogen and to contain two to three times more potassium than farmyard manure. These nutrients are released when the leaves decompose. Not surprisingly, comfrey has many uses, in particular in the form of a concentrate feed for plants. But take care – if you grow the wild form of comfrey it will set seed everywhere, and its deep roots make it a devil to remove. Instead, buy the sterile variety 'Bocking 14'. It establishes quickly, and you can cut it back to ground level several times during the summer months. All it needs by way of care is a bit of mulching or compost to replenish nitrogen in the soil around its roots. It can be propagated by root cuttings.

The spring growth of comfrey can be cut and will soon regrow.

Comfrey tea and liquid feed

For a plant feed, pull off some comfrey leaves and stuff them into a large plastic bottle that has had the bottom cut off. Remove the cap and turn the bottle upside down over another container. As the leaves rot, a very smelly black liquid oozes out into the container. Alternatively, fill a bucket with shredded leaves, pack down and add a weight (do not add water). After a short time, the black liquid will be found at the bottom of the bucket. This liquid has been found to be as effective as ready-made plant feeds. As it is very concentrated, dilute with water by 1:20 before using.

To make a comfrey tea, shred some leaves, place them in a bucket or plastic dustbin and add water. Leave it for a few weeks. The leaves rot, creating a smelly dark liquid which can be drawn off and also used as a plant feed. To replenish, simply add a bit more water and fresh leaves to the bucket, and you will have a constant supply of nutrients for your plants. You can also use the liquid as a foliar spray, and this is believed to help plants resist mildew. It helps to add a drop of detergent for a foliar spray, as this helps the tea stick to the leaves. You can even spray your apple trees with it, to help reduce the incidence of disease.

Often a lack of nitrogen slows down the breakdown process in a compost heap, and this can be overcome by adding an activator. Comfrey leaves make a pretty good compost activator, as they rot down quickly, releasing lots of nitrogen. When you have a plentiful supply of leaves you can use them as a thick mulch around plants such as tomatoes, where the potassium will be of benefit, or put them in a sack and allow them to rot down to form a leafmould, as you would fallen tree leaves in autumn. The leaves can also be used to line a bean trench, where they will rot down and release their nutrients.

Comfrey is also great for animals, as the leaves are about 25 per cent protein (much higher than that found in alfalfa or soya) and also low in fibre, making it suitable for feeding to chickens and pigs. Don't feed it fresh, as the animals will not like the prickly leaves; instead, let the leaves wilt for a few days or so first. You can also dry it and use it in poultry feed as a nutritional supplement.

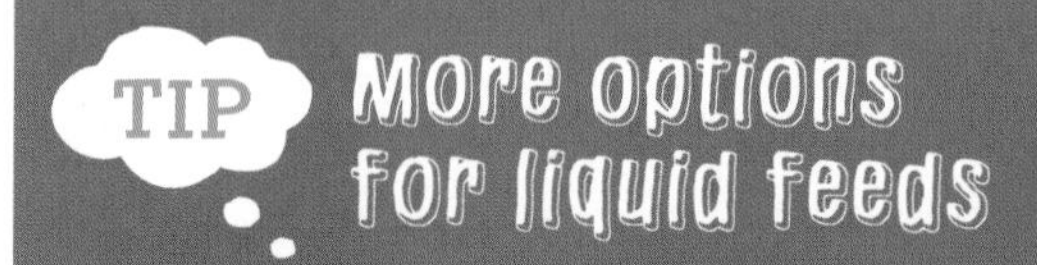

It's not just comfrey that makes a great liquid feed. Other plants have high levels of nutrients, for example nettles and ground elder. Their leaves can be seeped in water or even mixed in with comfrey and left for several weeks. The rotten gunge left at the bottom of the container can be added to the compost heap. Nettles are rich in nitrogen, iron, magnesium and sulphur, but do not have the levels of phosphate and potassium found in comfrey. Choose young plants that have not yet flowered.

Green manures

A green manure is a ground cover of mostly fast-growing plants, which is sown after a main crop has been harvested. The plants protect the soil and smother weeds. The green manure is either turned into the soil before the next crop is sown, so it can be broken down by soil decomposers, or it's chopped up and left on the surface to rot down, creating a mulch layer. Either way, a green manure helps to improve water retention and prevent the loss of nutrients by leaching, especially in the case of nitrogen and in particular on lighter soils, and it helps drainage on heavy clays. It also adds some nutrients and improves soil structure too, by adding bulk.

Green manures are cheap and easy to establish, and there are a number of plants that are suitable for the purpose. The choice depends on the time of year and whether the requirement is for protecting the soil, building fertility or both.

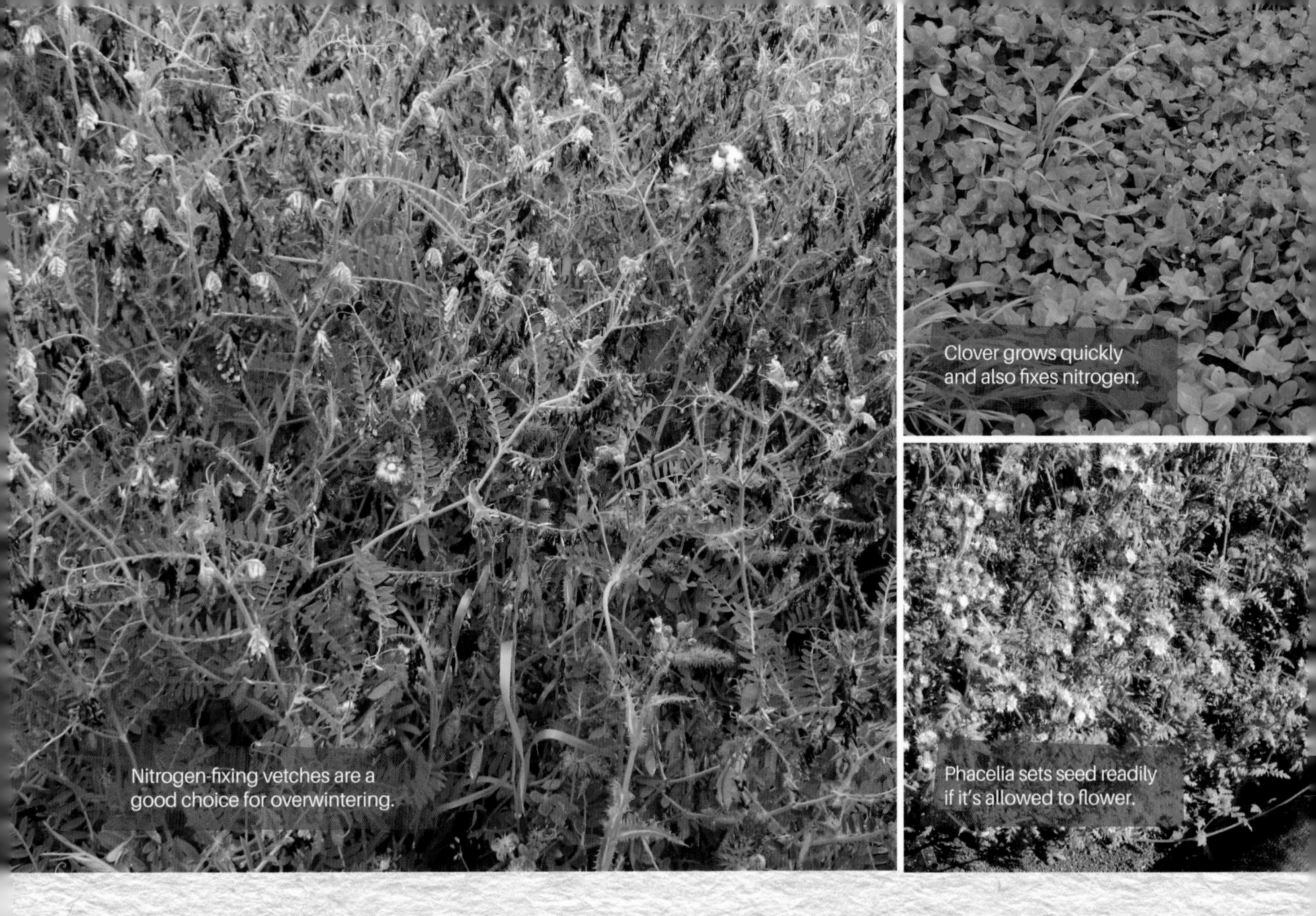

Nitrogen-fixing vetches are a good choice for overwintering.

Clover grows quickly and also fixes nitrogen.

Phacelia sets seed readily if it's allowed to flower.

A green manure helps to improve water retention and to prevent the loss of nutrients by leaching.

Some green manures, such as fenugreek, crimson clover and Persian clover, germinate and create a thick cover in a few months. They are used as a break in the crop rotation and, being nitrogen fixers (see below), they will boost nitrogen levels. Others, such ryegrass, phacelia and buckwheat, can be sown in autumn and left in the ground until the spring. These overwintered green manures are valuable in giving protection from nutrient loss, but they do not boost soil nitrogen significantly, even if they are nitrogen fixers, as the winter temperatures are too low for nitrogen fixation to take place (a minimum of 8°C/46°F is needed). Some green manures, for example lucerne (also called alfalfa) and chicory, have deep roots, so sowing these plants helps to extract nutrients from deep in the soil and to further improve the soil structure, especially on compacted soils.

One of the best ways of boosting soil nitrogen and improving soil structure at the same time

crop rotation helps to ensure that enough nutrients are available to the different crops each year.

is to grow a green manure of nitrogen-fixing legumes such as clover, vetch or lucerne. Legumes are a family of plants with small nodules on their roots which are filled with N-fixing bacteria: the bacteria take nitrogen from the air and convert it to nitrates that the plant can use. It's a mutual relationship, as the bacteria get sugars from the plant while the plant gains nitrogen. Fast-growing legumes, such as crimson and Persian clover, tend to be used in summer, when the temperatures are warm.

Crop rotations

A crop rotation, as the name suggests, is the practice of moving crops from one area to another each year, so they are not grown in the same soil for 2 years running. This helps to prevent the build-up of pests and disease. Crop rotation can also help to control weeds, maintain soil organic matter and ensure that enough nutrients are available to the different crops each year.

There are various rotation systems to choose from, depending on how much space you have and how long-term a rotation you want. They vary from just a couple of years to 8 or more years. Most organic growers opt for as long a cycle as possible, especially when growing onions, and also include a fertility-building crop such as a green manure in their rotations. However, rotations are only a guide and can be tweaked to suit your own needs, so don't feel that you have to stick to a rigid planting order. It is more important that you have a healthy soil with a rich diversity of soil life than that you slavishly follow a rotation plan. Indeed, some of the latest thinking is that crop rotations are not always essential: those following 'Natural Agriculture' systems such as Shumei, for example, will grow the same crops on the land with a no-dig approach for many years.

The main crop groups for a rotation are as follows:
Brassicas: Brussels sprouts, cabbage, cauliflower, kale, kohlrabi, oriental greens, radishes, swedes and turnips
Legumes: beans and peas
Alliums: garlic, leeks, onions, shallots
Potato family: potatoes, tomatoes (also aubergines and peppers, but they suffer from fewer diseases and can be used anywhere in the rotation)
Roots: beetroot, carrots, celeriac, celery, Florence fennel, parsley, parsnips and all other root crops.

When a crop rotation is used, crops that require the same soil treatment are kept together as much as possible, to ensure that they have the best possible growing conditions. This also makes the soil management more straightforward, as it means that over the course of the rotation a whole growing area receives the same treatment. For example, a well-rotted farmyard manure can be added to the soil prior to growing potatoes, leeks, brassicas and marrows, but not on beds designated for carrots, parsnips and beetroot, as the manure tends to be high in magnesium, which can encourage forked roots. It is also linked to white rot in alliums. Lime can be added to the soil before planting if necessary in order to maintain a neutral soil pH, for example added to the brassica beds the autumn before planting to help discourage club root, but it is not spread on potato beds as it can encourage scab. Brassicas are hungry crops, so planting them after a legume crop will mean there should be a good level of nitrogen present in the soil.

Crops such as cucurbits (courgettes, pumpkins, squash, marrows and cucumbers), French and runner beans, peppers, spinach, chard, salads (e.g. endive, lettuce and chicory) and sweetcorn do not suffer from soil-borne diseases, so can be dropped anywhere into the rotation, but it is good practice to avoid growing them too often in the same place.

Crop rotation also helps to suppress weeds, as some plants have dense foliage (courgettes, cabbage and lettuce, for example) and their shade keeps the weeds down. Other crops, such as onion and carrot, are more erect, so do not compete for light. Alternating plants with these different growth habits helps to keep weeds under control over successive years.

Some rotation options

A 4-year rotation is the most common:
Year 1 Potatoes/tomatoes plus squash, preceded by adding manure to the plot
Year 2 Legumes
Year 3 Brassicas, preceded by liming the soil
Year 4 Roots and alliums, followed by manure.

Year 1
Year 2
Year 3
Year 4

This ground has been cleared by pigs. Once they have gone, we rake over the soil as seen here, and cover it for the winter.

Ideally, an equal area is allotted to each group, but in reality there is no point putting aside a large area to grow a particular crop if you don't like it! Instead, you could keep to the general rotation plan but drop in a larger area of salad leaves or spinach. Also, changes may have to be made mid-season due to failures or problems with the weather. And, should disease crop up – in particular white rot in the allium family – you may need to extend the rotation.

Longer rotations give more balanced soil fertility, a greater gap between crops of the same family, and the opportunity to include green manures to help build fertility.

For example, an 8-year rotation might be:
Year 1 Potatoes
Year 2 Sweetcorn or green manure or mulch
Year 3 Brassicas
Year 4 Legumes
Year 5 Tomatoes
Year 6 Legumes
Year 7 Roots and alliums
Year 8 Squash.

Some crops are perennials and will have a permanent location. Examples include artichokes and asparagus (see Chapter 4, page 75).

Rotations with livestock

Ideally, you should aim to include some of your livestock in the rotation, to boost fertility. This can be achieved on larger plots rather than on small vegetable beds. For example, pigs will turn the soil of a plot, add nutrients in the form of manure, and generally clear the ground. Once the pigs are removed in autumn, any remaining weeds can be pulled out and the ground can be covered. It is best to use it for hungry crops such as potatoes the following year. After potatoes, the ground can be seeded with a grass mix for chickens. Chickens, too, will add fertility, and they will clear some of the vegetation by scratching the ground. Once the chickens are moved, check the soil pH, as chicken faeces can make the soil acidic, so you may need to add lime. The ground can then be used for crops such as squash, spinach, or even brassicas. Then the ground is cleared and re-seeded with a grass mix, ready for the return of the pigs – giving a 4-year cycle that breaks soil-borne disease in crops as well as parasitic disease in pigs and poultry.

Including green manures

Green manures should also be included in your crop rotations. For example, a white-clover cover can be sown under squash or brassicas. The clover will grow slowly during the summer, as it gets shaded by the leafy plants, but will persist. By the end of autumn the white clover will have spread out and created a good winter cover. The following spring, it can be dug in or left in place and the next crop planted into it.

Traditionally, a legume crop is included in the rotation after a hungry crop, such as potatoes,

Once the crops are harvested on my plot, the beds are sown with various green manures, which cover the soil over winter and for much of spring.

to boost nitrogen levels. But the benefits to the next crop come only when the above-ground residues of the legumes, together with the roots and nodules, are dug into the soil and decompose. If you are growing the legumes as a crop, you will be removing a lot of the nitrogen when you harvest the peas and beans, so any benefits are surprisingly small. This is why a nitrogen-fixing green manure should always be included in your rotation.

Here is an example of a 5-year rotation that makes good use of green manures to maintain fertility. As described below, green manures are integrated into the rotation at each stage of the cycle:

Year 1 Potatoes or sweetcorn
Year 2 Onions, leeks and garlic
Year 3 Brassicas
Year 4 Roots (carrots, beetroot) and legumes
Year 5 Clover

Here, the potatoes are cleared by autumn of the first year and the ground is sown with a rye and vetch mix. Sweetcorn can be undersown at the time of planting with low-growing trefoil or white clover, which remains through the winter. Any garlic and onions for overwintering can be planted in autumn too. Once the onions and garlic (or leeks) have been harvested, by autumn of the following year, a green manure of red or crimson clover can be sown, to remain until next spring when the brassicas are planted out. Any early brassicas can be followed by a summer green manure of crimson or red clover, buckwheat or phacelia, while the main-crop brassicas can be undersown with white clover. In year 4, carrots, beetroot and legumes are grown and, as they are harvested, the bed is sown with a mix of rye grass, clovers and vetches. This remains for 16 months until the next potatoes are planted. The year-4 legumes are undersown with white clover, which also remains through year 5.

PART TWO

GROWING PRODUCE

chapter FOUR

The vegetable & flower garden

It's surprising just how much can be grown in a small area, so there's no need to devote a large part of your plot to vegetables. There will be plenty to do elsewhere on the plot! Nor do you have to limit yourself to annuals: try growing perennial vegetables, and some flowers for cutting.

The traditional UK allotment is 250m² (around 300 square yards, or, in 'old money', 10 square rods, when the rod used to control a team of oxen was 5½ yards long). An allotment was deemed to be of sufficient area to feed a family of four for a year. You can fit 16 standard allotments into an acre! An acre – or even less – is still large enough to have plenty of space to rotate annual crops, grow some perennial crops, include some flowers and erect a greenhouse or polytunnel. That is a decent space!

The key to successful growing is to be able to take control of your plot and not feel that you are controlled by it. You need to be able to find time every week to keep on top of jobs: if you get behind for a few weeks, things will get badly out of control. It is far better to start with a smaller area and keep it looking good than to take on too much and get overwhelmed, so think carefully about the area you can realistically take on for annual crops.

In this chapter, I discuss planning your vegetable garden, weed control, companion planting and trying out perennial crops as opposed to annuals, plus sparing some space for flowers – not just for pleasure but also for other benefits. However, I am not giving detailed advice on growing specific crops, as there are plenty of other books that deal with this topic in depth: see Resources for some examples.

Visiting local allotments is a great way to get an idea of what crops to grow and how to arrange your plot. Every allotment on this site at Lytes Carey in Somerset has been laid out differently.

when deciding what vegetables to grow, always choose crops that you like to eat!

Planning your vegetable beds

What do you want to grow? The best place to start with a new plot is to make a wish list of all the vegetables you want to grow. Always choose crops you like to eat! I like to include unusual veg that I can't buy in the supermarket. Also, try to choose those that grow well in your area, so talk to local growers or visit local allotments. Try not to grow too much, otherwise you'll end up with a glut, although the pigs will always eat it! Successional sowing, i.e. sowing small amounts of seed every few weeks, will help to stagger the harvest so you can eat your produce at its best. Once you have your list, you can divide it into family groups and plan your crop rotation (see Chapter 3) and map out the growing area.

Bed size and style

In a traditional system, the whole of the vegetable area is cultivated, with dirt paths providing access between rows of crops. But more modern thinking is to opt for a bed width of 1.2-1.3m (4'-4'3") separated by paths, so you can reach across and never need to walk on the soil. Some prefer narrower beds, but remember that smaller beds mean proportionally more path, so more land is

The width of these raised bed is fine, but their long length may tempt you to take a short cut and step across the bed

lost. The beds can be level with the ground or raised, in which case they are surrounded by a frame to contain the soil. Whatever size you opt for, standardize this through the growing area. This way, any cloches, fleeces and supporting systems that you buy will fit any bed.

Orientation

The best orientation for beds is north to south, to minimize the amount of shading, with, where possible, the taller plants growing to the shaded end of the beds (in the northern hemisphere this is the north end). Perennial plants tend to be larger and should be positioned in spaces where they are not going to shade the annual crops, so this is usually around the edges of the plot.

Planting patterns

Traditionally, vegetables were planted relatively close together in rows, which suited the seed

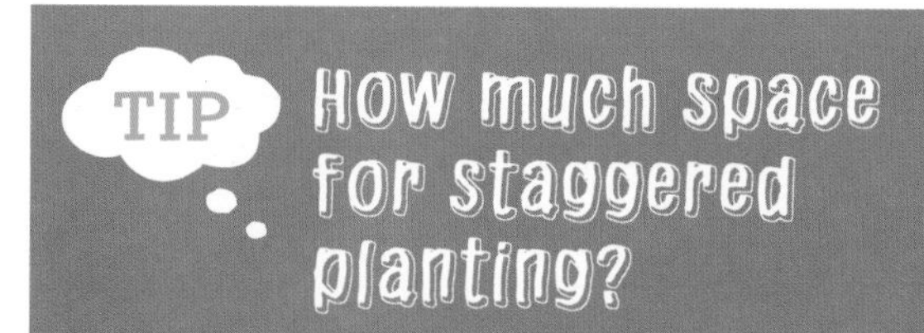

How much space for staggered planting?

As a simple rule of thumb, take the recommended spacing for within and between rows and take an average. For example, if the packet says thin to 8cm (3") and allow 20cm (8") between rows, for an equidistant spacing you want (8 + 20)/2 = 14cm. (Or (3 + 8)/2 = 5½".) If you want to grow small or baby vegetables, then you could opt for a closer planting, getting down to as many as 12 beetroot or carrot plants in a square foot (0.1m²).

Onions planted in rows . . .

. . . and shallots in a staggered pattern.

This spinach is growing between broccoli plants, which will mature much later in the year.

drill, and guidance on seed packets still talks about rows. There are advantages to having long rows: they make sense if you use a seed drill, and if you are digging a potato or bean trench, for example, then it's easier to have one long trench than four short ones. Many of the fleece and insect meshes come in relatively narrow widths, which suit rows rather than beds, and it's easier to hoe a row.

But, increasingly, research shows that you can boost productivity by abandoning the row and opting for a more even distribution across the bed. An even spacing means that there is less competition between roots or shading from neighbouring plants, and the plants make better use of the nutrients. The result: more productivity. I find that it does take more time to hoe around plants in a staggered, equidistant planting pattern than in a row, but the crops also create a weed-suppressing cover more quickly, so there are fewer and smaller weeds. I prefer rows for onions, beetroot and carrots, because of the ease of sowing and hoeing, but for crops such as brassicas, squash, broad beans and sweetcorn I use a staggered pattern.

Intercropping

Intercropping is the practice of making use of the empty space between crops when they are young by sowing or planting a quick-growing crop (also known as 'catch crops') – for example, growing radishes or lettuce between parsnips or brassicas. The main crop is slow to get established and will most likely have an erect habit

Don't let weeds set seed! The old adage 'One year of seed equals seven years of weeds' is so true.

and be widely spaced, as with leeks, brassicas, sweetcorn or peas. In contrast, the second crop is small and, because it is establishing while the main crop is still small, it is not shaded out. Good examples are lettuce, radishes, spring onions and beetroot grown for baby vegetables. Timing is critical with intercropping: get it wrong and the second filler crop may compete with the main crop, reducing your yield. Intercropping will take up more nutrients from the soil, so the beds need to be well composted to maximize the yield of both crops.

Controlling weeds

I have met so many new allotment holders or smallholders who have cleared a load of beds, sown their seeds and disappeared off for a few weeks, only to return to find the beds covered in annual weeds. It's a very depressing sight, and one that can easily put off the novice.

Weed control is critical, especially in late spring and early summer, when it seems that you are fighting a constant battle. If you can get through this stage it gets much easier, as the crops get larger and shade out the weeds, and the store of weed seeds at the surface of the soil has been exhausted. Persistent weeds such as thistles, docks, bindweed and ground elder need to be tackled firmly, otherwise the problem just gets bigger each year. If you are not sure of the identity of a weed, try downloading a weed app on to your smartphone, so you can look them up while on the plot.

For advice on clearing weeds and preparing the ground on an uncultivated or overgrown area of land, see Chapter 2, page 38.

Stale seedbeds

If you prepare your beds well, this will help reduce the number of weeds later in the season and save many hours of hand weeding. One way to do this is to create stale seedbeds. (This is particularly useful for growing carrots.) The soil is prepared to a fine tilth – the surface lumps broken up to a crumbly texture – and left for a couple of weeks. This allows the weed seeds in the soil to germinate, and you can then clear them with a hoe or a flame weeder, but remember to keep the hoeing shallow, as if you go too deep all you will do is bring new seeds to the surface. Then the soil is ready for sowing the seeds that you want to germinate!

Little and often

As many allotmenteers know, it is so frustrating to control the weeds on your own patch, only to see them creep in from neighbouring plots! I have received lots of advice regarding weeds over the years, but among the best is to carry a hoe around and take out the annual weeds as soon as they germinate. I have long used a push–pull hoe fitted with an extra-long handle

Erect crops such as parsnips and carrots can soon get overwhelmed by weeds, so need regular hoeing.

so I don't have to bend down. I now have two of them, one of which is left in the polytunnel so there is no excuse not to hoe the weeds when I go in to harvest.

Many years ago, a very experienced gardening friend told me to always return to the area where you last weeded, as it won't take long to clear it before moving to a new patch. Using this method has helped me get some new areas under control in my walled garden. And, most importantly, don't let weeds set seed, especially docks and thistles! The old adage 'One year of seed equals seven years of weeds' is so true. In our first smallholding, I decided that a patch of thistles should be left, as they attracted loads of pollinating insects. Big mistake! We battled thistles in that field for many years. Now we top our fields as soon as the flower heads appear, following the advice of Pauline Pears of what was then the Henry Doubleday Research Association, which was to leave the thistles until the middle of summer, when they were at their tallest and in flower, and then cut them to the ground to weaken them.

We carry a 'jungle blade' when we walk the dogs and chop off weeds as we go, and it's surprising how much clearance can be done by just doing a little bit every day. In mid spring on my walks I focus on pulling out docks by hand, as the soil is moist and I can get the taproots out. My target of pulling 20 clumps a day has resulted in several fields being completely cleared of docks.

A raised bed covered with biodegradable mulch that slowly breaks down over summer.

A ground cover also helps to stop mud splashing on to salad leaves.

These young brassica plants have been undersown with nasturtium, creating a living mulch to suppress weeds. It will remain once the crop has been harvested. The nasturtiums lure cabbage white butterflies away from the main crop.

Weed barriers

Many annual seeds need light to germinate, so covering the ground with a mulch to prevent sunlight reaching the soil will stop many weed seeds germinating. Also, the mulch traps moisture, creating a microclimate that encourages root growth and helps transplants to establish. A mulch is simply a top dressing, and can take many forms. It can be a covering of straw, compost or bark, or sheeting of some sort.

A popular choice is the black woven ground-cover fabric, which comes in different widths. It's not cheap, but it lasts a number of years, and it is permeable – suppressing weeds while allowing water to pass through, so it's a real time saver. You can lay the fabric over the ground and plant through it. We use it a lot, both on our paths and on larger beds where we know we won't be able to control the weeds. For example, we use it to cover the pens cleared by the pigs as soon as the pigs are moved off.

You can get biodegradable mulch now, either paper or a plastic film. It is permeable and lasts about 16 weeks, slowly rotting down over the season and eventually disappearing altogether. Again, it is not a cheap option, but it does suppress the weeds and saves time.

Welsh onions in flower. Remove the flower heads from young plants to encourage root growth.

This Swiss chard plant has been growing in my polytunnel for 3 years. It is cut back in spring to prevent flowering.

Another option is the 'living mulch'. This is a ground cover of plants growing under the main crop, for example a low-growing green manure such as white clover (see Chapter 3, page 62), or nasturtiums growing under brassicas. The living mulch suppresses the weeds while the main crop gets established. Care has to be taken with this option that the mulch does not grow too vigorously and compete with the crop plants.

Perennial vegetables

Perennial vegetables are a really useful addition to the plot. Some live a few years; others many decades. Some persist year-round while others die back in autumn and re-shoot in spring. Say 'perennial vegetable' and you probably think of artichokes and asparagus, but there are many more to choose from, offering a year-round supply of leaves, shoots, flowers, fruits or roots.

The advantages of growing perennial vegetables is that once they are established you don't have to do much to keep the supply of edibles coming, other than a bit of weeding. Many provide tubers or leaves during the winter months, when other crops are in short supply. They don't have to take up space in the vegetable beds, but can occupy those tricky spaces and corners. Some of the larger ones, such as artichokes and perennial kale, can be grown as barriers and hedges within the plot. Not only do perennials supply food, but they also enhance the plot in other ways, for example by attracting pollinating insects and providing a habitat for insect predators. Some are leguminous and will fix nitrogen, while others provide shade. They improve the soil too: unlike with annuals, there is no disturbance of the soil, and each year their root system gets larger, boosting the soil food web, while their fallen leaves add to the leaf litter and thus to the soil organic matter. Nutritionally they are better too, as their deep roots draw up nutrients from the subsoil in the same way as comfrey, as described in the last chapter (see page 57).

Some annual crops can actually be grown as perennials. In my polytunnel I have a Swiss chard plant that has been going strong for 3 years and now sports a huge woody stem. I don't let the plant set seed. Instead, I cut it back in late spring, give it a good feed and let it recover to give me a supply of new leaves all winter. There is a vast range of perennials to choose from, and more are becoming available to the grower each year. In the table on the next few pages I describe some of the more commonly grown perennials, which I consider worthy of space in your plot.

A selection of perennial vegetables

Common name(s)	Latin name	Characteristics and cultivation advice
Asparagus	*Asparagus officinalis*	Buy one-year-old crowns, dig a trench and backfill with compost. Then form a small mound along the trench and plant the crowns so they sit on the mound, then backfill. Allow 70cm (28") between rows and 30cm (12") between plants. Let the crowns establish for a couple of years before harvesting. Don't waste the space between the rows – plant them with other perennials such as chicory.
Babington's leek	*Allium ampeloprasum* var. *babingtonii*	A wild leek with thin leek-like stems and a mild garlic-like flavour. It doesn't set seed, so you need to plant bulbils in a sunny or part-shady position, and allow them to grow for a year before harvesting. The stems are cut at ground level in late autumn to winter and the underground bulb re-shoots in spring. It can reach up to 2m (6'6").
Buckler-leaf sorrel (French sorrel)	*Rumex scutatus*	A low-growing compact plant with shield-shaped leaves. Easy to establish from seed but it does spread readily! The young leaves have a sharp flavour and can be used in salads or wilted and used with meat and fish.
Cardoon (artichoke thistle)	*Cynara cardunculus*	This large thistle-like plant is actually related to the sunflower and was a firm favourite of the Victorians. Research has shown that the cardoon and globe artichoke are descended from the eastern wild cardoon; the Romans domesticated the globe artichoke, while the cardoon still resembles the wild plant. Cardoons are grown for their flowers, which attract bees. They are edible in part: the leaf stalks can be harvested in winter, blanched and used in gratins.
Chicory (leaf chicory, Italian chicory, radicchio)	*Cichorium intybus*	A herbaceous perennial with blue flowers. It can be grown as a forage plant for livestock, but is more usually grown for its leaves, buds and roots. It is a useful plant, as its deep roots bring up minerals from deep in the ground. The leaves are used in salads while the roots can be used as a coffee substitute. It can be established from seed, planted out at a spacing of 20cm (8"). Plants need to be well watered in hot periods to prevent the leaves becoming bitter. The leaves are harvested by cutting them close to the base of the plant. Chicory is a hardy plant that can be sown late and harvested through winter. Plants can be forced by growing them in a dark place such as a cellar.
Garlic	*Allium sativum*	Garlic is usually grown as an annual, with the bulb planted in winter and harvested in summer. However, it can be grown like perennials. Simply leave some of the bulbs in the ground and they will come up again the following spring. Garlic likes full sun, and although it will grow well in most soils, it does not like heavy clay soils prone to waterlogging. I have a clump of garlic plants in the polytunnel that has survived for 5 years, and each year I harvest a few cloves. They self-seed too, providing me with seedlings for salads or for transplanting.

Common name(s)	Latin name	Characteristics and cultivation advice
Giant butterbur (bog rhubarb, fuki, Japanese butterbur)	*Petasites japonicus*	A large plant with huge rounded leaves that resemble rhubarb. It needs boggy ground and is tolerant of shade, so it can be grown under trees, but it can be invasive. The edible stalks are eaten in the same way as rhubarb.
Good King Henry (poor man's asparagus)	*Chenopodium bonus-henricus* (also known as *Blitum bonus-henricus)*	A useful, unfussy plant, growing to 60cm (2'). The leaves can be continually harvested, even through winter. They are bitter, so use salt to draw this out before cooking. Harvest the young spring shoots and cook like asparagus, and use the flower buds like broccoli.
Globe artichoke	*Cynara cardunculus* Scolymus Group	These big thistle-like plants are grown for their large flower buds, which are harvested before they open. Sow seeds in spring and transplant the young plants at a spacing of 60-90cm (24-36"). They grow well for a couple of years but then need to be divided to retain their vigour. In winter, mulch with straw or well-rotted manure to protect from frost.
Horseradish	*Armoracia rusticana*	Growers have a love–hate relationship with this plant! Be careful where you plant it, as once it's in the ground its roots take hold and it's almost impossible to get rid of. It is grown for its roots, which have a strong mustard flavour, but the leaves are edible too. Plant by digging a hole and backfill with compost. Place a root into the hole at an angle, cover with soil and then water.
Lovage	*Levisticum officinale*	A tall perennial herb, growing to 2m (6'6") or more with long, hollow stems. The young leaves can be used as a substitute for parsley or celery, but it's got a more punchy flavour. It can be shredded and added to stews or scrambled eggs. It has medicinal value too. The large leaves die back and are replaced, so by clipping you get a steady supply of fresh leaves. The new stems are tender and can be steamed and served with chicken, while the taproots can be peeled and used like salsify. Harvest the seeds and substitute them for celery seed in pickles.
Perennial kale (everlasting kale)	*Brassica oleracea* Ramosa Group	There are various perennial kales, including my favourite, the Taunton Deane kale or cottager's kale (pictured on page 80). This one grows to 1m (3') or more in height, growing for years to yield a supply of greens year-round. The deep roots bring up minerals from the subsoil, so the leaves are particularly nutritious. It's an unusual plant as it doesn't set seed, so it has to be propagated by taking cuttings. Simply take off the side shoots and pop into potting compost. A close relative is Daubenton's perennial kale, which grows to about 70cm (28").
Rhubarb (Victoria rhubarb)	*Rheum* x *hybridum* (also known as *Rheum rhabarbarum*)	A familiar herbaceous plant with large leaves and fleshy-coloured petioles (stalks) that grow from thick rhizomes. It has a distinctive tart taste and is commonly used as a fruit, although it is a vegetable. It is an easy plant to grow and prefers free-draining rich soils in partial shade. It is usually established from crowns (rhizomes) in autumn and winter, at a spacing of 80-120cm (32-48") between plants. In summer, once the leaves have died back, mulch around the plant to conserve moisture and keep the weeds down, but do not cover the crown.

Common name(s)	Latin name	Characteristics and cultivation advice
Salad burnet	*Sanguisorba minor*	An attractive clump-forming perennial with saw-edged leaves that have a cucumber-like flavour. It is also used as a medicinal herb. Can be grown from seed.
Sea kale	*Crambe maritima*	A maritime plant that grows wild along coasts, found in shingle banks. Another Victorian favourite, it is grown for its forced white stems, which are eaten from the middle of winter to early spring. The leaves can be harvested too and fried in olive oil. Sea kale can be established from seed in spring. It prefers a rich, well-drained soil in partial shade.
Skirret (perennial parsnip)	*Sium sisarum*	This was once eaten as a root across Europe in the Middle Ages, but it lost favour once the potato arrived. The skinny roots taste of mostly parsnip and carrot, with a hint of potato, and can be harvested all year. Grow from seed in spring and plant out in well-manured beds. By late summer the plants reach 1.5m (5') in height and have lots of small white flowers. It is best to allow them to grow in the first year and get established before taking a crop. The following year you will have lots of thin roots. Keep the plants well watered, otherwise the roots can be tough. You can propagate by dividing the roots.
Sweet cicely	*Myrrhis odorata*	This herbaceous perennial can grow to 2m (6'6") or more. It has feathery leaves and umbels of white flowers. The leaves have a strong smell of aniseed and are used as a herb. The roots and seeds are also edible.
Walking onion (Egyptian onion, tree onion)	*Allium cepa* Proliferum Group	This onion has chive-like leaves, which can be used like spring onions. Allow some to grow bulbils on the end of their leaves, which will bend to the ground and root, hence the term 'walking'.
Welsh onion	*Allium fistulosum*	An evergreen, hardy, perennial onion that is ornamental as well as culinary. Despite its name, it originates from China. These green onions form bulbs similar to those of spring onions. The shoots die down in winter. They can be established from seed.
Winter savory	*Satureja montana*	An aromatic herb with dark green, narrow leaves and white flowers. It is grown as a companion plant around beans, as it deters mildew, aphids and weevils. It grows to about 50cm (20") and is a good border alternative to the dwarf box.

This striking plant is tree spinach (*Chenopodium giganteum*), a relative of Good King Henry. It is an annual, growing to 1.5m (5'), and can be used for salad leaves and as a green manure.

The towering Taunton Deane kale, also known as cottager's kale, is one of my favourite perennial vegetables.

These Jerusalem artichokes form a summer barrier in the vegetable plot.

Tubers and roots

Although not strictly perennials, there is a whole host of tubers that are useful and low maintenance. Many of these are from South America, the home of the potato.

The **Chinese artichoke (*Stachys affinis*),** despite its name, is not related to the globe artichoke (nor to the Jerusalem artichoke – see below). This is a much smaller plant than either, growing to about 60-70cm (24-28"). The tubers are tiny, crunchy with a nutty flavour, and can be eaten raw in salad or stir-fried. Plant out the tubers over winter and harvest in autumn.

The **Jerusalem artichoke (*Helianthus tuberosus*)** is among the most familiar of edible tubers. A relative of the sunflower, it grows up to 2.5m (8') in height, and a row of these plants can form a useful and attractive barrier on a windy site. The flowers are attractive to bees too. The tubers are dug up in autumn, the best selected for the next year's crop and replanted, and the rest eaten. They are prone to falling over during windy weather towards the end of summer, so stake them before they get too large. Cut off the dying stems and leave the tubers in the ground until needed, as they do not store well.

Oca (*Oxalis tuberosa*) produces a small tuber that comes in various colours and is used just like a potato. The advantage of this easily grown 'Inca' crop is that it is blight-free. The tubers produce bright green clover-like leaves and attractive yellow flowers. However, the

Yacon has distinctive large leaves. Underground are the edible storage tubers and the smaller growing tips that are replanted in spring.

Many useful plants with edible tubers come from South America, the home of the potato.

tubers only start to form in autumn and are harvested from mid winter to early spring. As the plants can be damaged by frost, care needs to be taken regarding where they are grown. The beds should be covered with fleece as soon as there is risk of frost, to extend the period of growth. Once the plants have completely died back, leave them for at least a week before you harvest, as the plant will continue to move food into the tubers as the stems die back. The tubers have a lemony taste and are cooked, unpeeled, like a potato. They suffer from few pests, although rabbits have a taste for the young growth.

Yacon (*Smallanthus sonchifolius*) is another South American plant. It grows to about 1m (3') high, with large, sunflower-like leaves. Plant out in spring after the risk of frost has passed, and harvest in early winter before any hard frosts. It produces two types of underground structures: large storage tubers that are very crisp and can be used raw in salads, and knobbly growing tips, which are removed along with the crown and overwintered in a frost-free place, rather like dahlia tubers.

More exotic choices include the **Chilean ulluco (*Ullucus tuberosus*)** and the **Peruvian ground apple or mashua (*Tropaeolum tuberosum*).** The Chilean ulluco produces attractive multi-coloured tubers that resemble sweets. These are planted in early spring and will create a low-growing spreading plant. Tubers are harvested in mid to late winter. They are susceptible to slugs and snails. The Peruvian ground apple is a South American perennial dating back 5,000 years or more, grown for its peppery tubers. It is easy to grow, and, though the shoot is not frost hardy, the tubers can survive in the ground to about -7°C (19°F). The high concentration of mustard oils in the tubers may account for its resistance to nematodes and other tuber pests. In its native area it is used as a companion plant to potato (see below), repelling the pests of the potato tubers, and has been found to have medicinal qualities too.

Companion planting

Companion planting is a form of polyculture – which, as the name suggests, means growing two or more plants together. Companion planting has a number of benefits, including deterring pests and disease, attracting predators or pollinators, or providing either nutrients, support or shelter.

Some companion plants are beneficial because of their strong smell, which either repels a pest or lures it away (this is often described as a push–pull strategy in farming). For example, leeks are often damaged by thrips, tiny insects that feed on plant juices, causing white spots on the leaves. This is a tricky pest to control, as it hides between the sheaths of the leek leaves. Planting a row of lavender plants near the vegetable beds has been found to lure the thrips away from the leek. Brassicas can gain protection from pungent plants such as the tobacco plant (*Nicotiana tabacum*), *Artemisia* species and hyssop (*Hyssopus officinalis*), as the female cabbage white butterfly finds the leaves of brassicas through smell.

Gardening books have long recommended that carrots should be grown with members of the onion family, as the pungent smell of onion helps to confuse the carrot root fly, which finds carrots by smell, but trials by Charles Dowding and other gardeners have found no benefit. It is better to net the plants or, since the female carrot flies are low-flying, surround the plants

Plants that attract pollinating insects are particularly beneficial, especially if you are growing fruit trees. Some partnerships to consider include a row of French marigolds by runner beans; sweet peas (*Lathyrus odoratus*) close to courgettes; and lavender planted near fruit trees. Remember to draw pollinators into your polytunnel to pollinate your early crops such as courgettes and broad beans, by planting marigolds (French or calendula) and early-flowering salvias.

with 60cm (2')-high barriers to keep the egg-laying females away.

Tall plants such as sweetcorn and sunflowers (*Helianthus annuus*) can provide shade for spinach and lettuce, or support for climbing French beans, while deep-rooted plants such as comfrey draw up minerals from deep in the ground, which are then incorporated into the topsoil through their leaf litter, to the benefit of plants growing around them. Equally, there are some pairings which should be avoided, such as legumes and onions. Onions exude substances from their roots which adversely affect the nitrogen-fixing bacteria found in root nodules of legumes.

The classic 'three sisters' combination of sweetcorn, squash and French beans.

A row of French marigolds has been sown beside these runner beans to attract pollinators and predators of pests.

French marigolds are key companion plants.

The following are some of the most important companion plants, and should be included in the plot wherever possible.

Basil is often growing in polytunnels alongside tomatoes. There is conflicting opinion as to whether its pungent smell repels or lures white-fly and aphids, but most are in agreement that this is a beneficial partnership. Basil is also believed to boost the growth of peppers and aubergines.

Bee balm or bergamot (*Monarda didyma*) can be grown near tomatoes to improve their growth and flavour. It is attractive to pollinators.

Calendula or pot marigolds (*Calendula officinalis*) have bright flowers that attract many predatory insects, such as hoverflies and lacewings.

Chives have leaves with a pungent smell from the presence of sulphur-based compounds, and this deters many insect pests.

Dill has yellow flowers that attract hoverflies, and also parasitic wasps that parasitize the larvae of cabbage pests.

French marigolds (*Tagetes patula*) are essential plants in the veg plot and polytunnel. Their pungent smell deters whitefly and aphids, benefitting tomatoes. Their roots exude sub-stances that help to control harmful nema-todes in the soil around roots of melons and tomatoes, and act against eelworms, which damage potatoes.

Mexican marigolds (*Tagetes minuta*) are useful plants, but they are tall. Their main use is as a bed clearer, as their roots exude an allopathic

Bug Housing

Another way to attract beneficial insects and other minibeasts into the plot is to build a bug house. This can be made from structures such as old pallets, stuffed with all sorts of materials such as bamboo canes, broken pipework, tiles, clay pots or fleece.

Place your bug house in a sheltered and sunny spot, out of the prevailing wind.

Brassicas can gain protection from pungent plants, as the cabbage white butterfly finds brassicas by smell.

substance which inhibits the roots of other plants. Mexican marigold has been found to be effective in clearing ground of pernicious weeds such as ground elder and bindweed. The roots are said to have insecticidal, nematicidal, antibacterial and fungicidal effects.

Mint is said to improve the growth of tomatoes and cabbage. There are various mints, and their aromas deter cabbage white butterflies, ants, flea beetles and aphids. Mint is an invasive plant, so it might be grown in its own patch and the leaves harvested and used as a mulch around vegetable plants. The flowers of the mint attract predatory insects such as hoverflies.

Nasturtiums (*Tropaeolum majus*) are grown as a lure for white butterflies, drawing them away from brassicas. The plants secrete a mustard oil that attracts the insects. Nasturtiums will also draw blackfly away from broad beans and French beans, and if grown in the polytunnel they lure insect pests away from tomatoes and cucumbers. Nasturtiums planted around the base of fruit trees also help to repel bugs such as aphids and scale insects.

Poached egg plants (*Limnanthes douglassi*) produce masses of open flowers that attract predatory insects such as ladybirds and hoverflies. Grow them around broad beans and other legumes to give them protection from aphids.

Radish can be grown between rows of brassicas and spinach to lure away the flea beetle. You still get a crop of radishes, as their roots are unaffected by the beetle.

The warmth of the polytunnel allows for early sowing of half-hardy crops.

In late winter I empty the plastic compost bin and fill it with fresh manure from the chicken houses. Temperatures inside the bin soon soar to 40°C (104°F).

Extending the season

In a temperate climate, you can gain many extra weeks of harvesting by means of a polytunnel or greenhouse, cold frames or hot beds.

All-year crops in the polytunnel

Because of its large size, a polytunnel can provide you with a whole winter growing season, so you can have salads and other crops all year round. You can keep crops such as brassicas, spinach, Swiss chard and beets going through the winter, and also get an early start with seeds in late winter and early spring. But polytunnels are not frost-free environments, so many growers put up a secondary cloche of bubble wrap inside for sensitive plants, or block off a section of the tunnel which can then be heated. Heating can be expensive, but you can help to keep the ambient temperature above freezing by filling large plastic containers or dustbins with water, to act as heat stores. Or fill them with fresh manure and green-waste materials, which will generate valuable heat as they decompose.

See Chapter 1 (page 28) for a discussion of the options when setting up a new polytunnel.

Polytunnel planning

Just as with your vegetable beds, it's important to plan the layout of your polytunnel to make best use of the space. With wider tunnels you may want two paths running the length of the tunnel, each wide enough for a wheelbarrow. I cover my paths with ground-cover fabric, but you could just as easily have an uncovered path. You can build raised beds too.

Rotating the crops in a confined space is tricky but still worthwhile. Try to maintain some rotation, however short the cycle: for example, move your tomato beds each year, and try to avoid growing the same crop on the same ground for 2 years running. The soil will soon get depleted of nutrients, so include fertility-building legumes in your rotation, such as a spring crop of broad beans or an autumn crop of peas, and if

Biodegradable mulches, both plastic and paper, cover the beds to suppress weeds over summer.

Try to rotate your tomato beds each year.

you have any space, sow a quick-growing green manure (see Chapter 3, page 59). Bring in compost regularly, and spread it over cleared beds.

Companion plants, especially calendula, French marigolds and basil, are invaluable for deterring pests and attracting beneficial insects. Remember, if you use insect screens on your polytunnel doors or raised sides, that you don't want to block out all the insects, as then you won't get efficient pollination.

Think about the position of the water supply, and the layout of your soaker hoses if you are using them, and make use of all the space by adding hanging bars and multi-tiered staging. And it is important to try to avoid using your polytunnel like a shed – a repository for all those pots, trays and bags of compost – as these harbour pests and disease, and also provide cover for the odd mouse. Although it can be very unpleasant working inside a polytunnel on a hot summer's day, make sure you pay attention to watering, as plants can get stressed quickly in high temperatures. You need to ventilate the

TIP Keep sowing seeds

You can sow crops in the polytunnel virtually every month of the year, so make a plan, as it's easy to forget to sow the winter vegetables in July, or to fall behind when you are busy sowing for your outdoor beds. With polytunnels, it's particularly worthwhile sowing little and often, to keep the harvests coming and avoid a glut.

Cleaning up with poultry

Pests can overwinter in a polytunnel and emerge ready to attack the new plants in spring. One way to reduce the overwintering of pests is to house your poultry in the polytunnel over winter. One cold winter I decided that one group of hens needed more shelter, so I popped them in the polytunnel. Although the soil was bare, it was not muddy. The chickens pecked over the ground and fertilized it for me, and I was delighted with the performance of that bed the following year. Quail are great for polytunnels too.

Cold frames can be used to harden off seedlings, to grow early crops, or to raise annuals for the flowerbeds.

tunnel well and remove diseased leaves as soon as they are spotted, as disease spreads rapidly in these conditions.

Cold frames

A cold frame is simply a structure with a clear top that lets in light and traps heat. Its main function is to raise seedlings and young plants in trays or pots and to harden them off before they are planted out. However, large cold frames can be used as mini-glasshouses, for example to grow plants such as melons, which require extra heat, or to get squash plants off to an early start.

Cold frames can be made from a variety of materials, but the key element is the transparent lid, which might be glass, polyethylene, fibreglass, for example. Some people use old window frames. Temporary cold frames can be

Move young plants into cold frames to free up space in your polytunnel or greenhouse.

This cloche, made from hoops of plastic pipe covered by a lightweight mesh, shelters young brassicas in spring and protects them from pests later in the year.

made from old straw bales, while permanent ones could be constructed from bricks, concrete blocks, plastic sheeting or planks of wood. Ideally, the sides of the frame are sloping, so that the cold frame catches more light.

You can dig out the soil below the frame to a depth of about 15-20cm (6-8"), so the surrounding soil provides insulation, although this can make the cold frame prone to flooding when it rains heavily. In late winter, a dug-out cold frame can be backfilled with fresh manure to create a hot bed (see page 90).

The cold frame needs to catch as much light as possible, so orientate it to the south (in the northern hemisphere), but make sure it is well ventilated, especially on sunny days, as heat builds up quickly and can soon desiccate young plants, and the extra humidity will increase the risk of fungal diseases. Prop open the lid up in the morning, but don't leave it too late in the day to close it, as you want to trap some heat to keep the temperatures up overnight. In spring, cold frames can be used to harden off seedlings started in the polytunnel or greenhouse, to grow early crops such as salad leaves, and to raise annuals for the flowerbeds. In autumn, you can sow crops that will benefit from the extra protection and continue to grow through winter, such as sweet peas and broad beans.

Cloches

Traditionally, a cloche was a bell-shaped jar used to protect a plant in its final growing position from frost. Now there is an array of cloches available, from simple plastic milk bottles to mini-polytunnels made from hoops and plastic, to give protection to more plants.

Cloches allow you to plant into the ground early in the season to get a head start. A courgette plant, for example, can be put out in early spring and given protection from frosts until the risk of frost has passed. Larger structures can cover whole rows of crops, such as an early carrot, beetroot or salad crop.

Hot beds

When manure and other organic materials rot down they generate heat, and this can be used to good effect in a hot bed. Hot beds are not new: the Victorians had many ingenious ways of heating their glasshouses, and the hot bed was one of them. Hot beds can be constructed within polytunnels to provide heat for a seed propagation bed or for early crops, but I prefer to build one outside, where I can sow early carrots and beetroot. Then later in the year I plant hungry crops like squash.

I make my hot bed from pallets, tying them into a square with string or plastic ties so the thing is easy to dismantle at the end of the season. I like to line mine with black plastic, to retain heat and make it easy to drag the compost on to the nearby beds at the end of the season, but a liner is not essential. I make a few holes at the bottom for drainage and then I fill it with fresh manure, stamping down between loads to compact it. Once it's filled almost to the top, I water it to kick-start the decomposition, add about 20cm (8") of compost on top and place a cold frame on top of that. Then I pack some extra hay around the cold frame to provide insulation. After a few days the temperature starts to rise and the bed is ready to plant up within the cold frame. The layer of compost is essential, as the manure will be far too hot for plant roots to withstand.

This hot bed was made in February with old hay and fresh manure from the poultry pens. The first lettuces were harvested in late spring, followed by carrots and beetroot.

The cut-flower patch

Most of the flowers sold in the UK and in other temperate locations have been grown overseas, often in Africa or South America. Doused in a cocktail of chemicals to make them last longer, they are flown across the world, clocking up thousands of 'flower miles'. In recent years there has been a surge in interest in home-grown flowers, raised without the help of pesticides or fertilizers and with little or no need for preservatives, using just water to keep them fresh. A cut-flower garden is perfect for a small plot, as the space needed is relatively small and the crop highly productive. My own 4m x 4m (13' x 13') cutting plot produces a large bucket of flowers every week from the middle of summer to early autumn, providing me with plenty to decorate the house and to give away. Flower farming offers much potential for a specialist business too. An area of an acre or less filled with well-grown flowers can generate a healthy income.

If you are well organized with successional sowings, you can keep your flowers performing well for many months.

What to grow

A look at any seed supplier's catalogue or website will show just what a wealth of flowers there is to choose from. With careful planning it's possible to have something to cut year-round. Choose flowers that grow well in your locality, and those with suitably long stems. Hardy and half-hardy annuals are a good place to start, as they can be sown in early spring and planted out after the risk of frost has passed. Most are easy to grow and will flower within 10 weeks. Further sowings can be made in late summer and autumn to get an early start the next spring.

Some species prefer to be sown direct, but in most cases transplants tend to do better than direct sowings. If you are well organized with successional sowings, it's possible to keep your flowers performing well over many months. These are some of my favourites for growing from seed:

Amaranthus (*Amaranthus caudatus*) comes in green and dark magenta forms, with catkin-like trailing flower heads.

Ammi majus looks a bit like cow's parsley and is used as a filler. Sow the seeds in autumn to get a good start, then sow successionally through spring. ***Orlaya grandiflora*** is similar but has slightly larger flowers.

Bells of Ireland (*Moluccella laevis*) can be tricky to get to germinate, but it has lovely shades of green and unusual flower spikes.

Clary sage (*Salvia viridis*) grows quickly, creating useful spikes of colour.

Cornflowers (*Centaurea cyanus*) should be sown successionally to keep the flowers coming.

Cosmos (*Cosmos bipinnatus*) for the prolific white flower heads.

Euphorbia oblongata has acid-green flowers that are great as a filler. It can be grown as an annual or short-lived perennial.

Honeywort (*Cerinthe major*) is one of my particular favourites, with nodding spikes of tubular flowers in beautiful tones of purple-and-silvery-green foliage.

Mexican sunflowers (*Tithonia rotundifolia*) are tall plants with vibrant orange flowers.

Pot marigolds (*Calendula officinalis*) are useful in several ways, giving early colour in the garden, as companion plants to attract pollinators, and as cut flowers.

Sunflowers (*Helianthus annuus*) come in various colours and heights. Allow a few flowers to go to seed and then watch the chickens squabble over the treat.

Sweet peas (*Lathyrus odoratus*) can be sown successionally under cover from mid autumn to mid spring for a summer-long crop.

Zinnias (*Zinnia* spp.) have lovely flower heads in a variety of colours. They flower until the first frosts.

My cutting garden, with a selection including pot marigolds, cornflowers, amaranthus, *Ammi majus*, cosmos and *Euphorbia oblongata*.

Remember to include the annual cutting patch in your crop rotation to avoid the build-up of pests and disease. Ornamental flowers suffer from relatively few problems, and they act as a useful break for your other crops.

Perennial flowers too can be rewarding to grow. Once established, they are easy to maintain, especially if planted through ground-cover fabric to keep down the weeds. Consider plants such as *Delphinium*, *Campanula*, *Rudbeckia*, *Echinacea*, *Eryngium*, *Phlomis* and *Verbena bonariensis*. Dahlias are back in fashion too. These showy plants are established from tubers, which can be overwintered. Climbing roses can be planted over arches to give colour to the garden and blooms for the vase. Choose one of the modern repeat scented flowering varieties to ensure a summer-long sequence of flowers.

Another essential for the vase is foliage, so think about plants that can be used for foliage or stems, such as dogwood (*Cornus* spp.), willow, rosemary and ornamental grasses. Some growers are focusing on specialist 'woodies' with interesting bark and leaves. These are trees and shrubs that grow back each year after cutting and provide useful fillers for the florist. Surprisingly, they can be in short supply, so it's a useful niche for a small plot, and this way you can extend the cutting time from late winter through to early the following winter. Don't forget fruits, berries and nuts from edible hedgerows, as well as attractive seedheads to give interest in autumn and winter.

Flowers for the plate

Still on the flower theme, another area that is growing in popularity, for both home use and as an income-earner, is edible flowers. It's amazing how a few flowers can liven up a salad, both visually and taste-wise. The range of flowers suitable for the plate is surprisingly large: borage (*Borago officinalis*), chives, cornflowers, courgette flowers, cowslips (*Primula veris*), French marigolds (*Tagetes patula*), lavender, bee balm (*Monarda didyma*), nasturtiums (*Tropaeolum majus*), pansies and violas (*Viola* spp.), peas, pot marigolds and wild primrose (*Primula vulgaris*) to name a few. Even the flowers of kale and other brassicas can be put to good use, as can the flowers of radishes and runner beans. In fact, many vegetables will provide a double crop of vegetable and flower. As with the cut flowers, with careful planning it is possible to have a steady supply of edible flowers available year-round.

Edible flowers have multiple culinary uses. For example, they add interest to salads, give flavour and colour to cakes and desserts, can be added to dressings such as vinaigrette, and used to make cordials, vinegar infusions, cake decorations and even food colouring. Among the most versatile is borage, which can be crystallized, added to summer drinks or frozen in ice cubes, and of course it attracts bees all summer long.

Most edible flowers are annuals that can be grown from seed in spring, either direct-sown or in modules to transplant. Others can be sown in late summer and early autumn to overwinter and provide an early supply in spring – for example, cornflowers, pot marigolds and violas. In spring, you can let a few brassicas run to seed to provide a supply of sweet-flavoured yellow flowers that are said to be highly nutritious. The key to successful edible-flower cultivation is to keep the flowers perfect and clean, as mud-splattered flowers cannot be used. For this reason, many of the commercial edible-flower growers use polytunnels, where their crops have protection from the elements.

There are additional benefits from growing edible flowers, as many are visited by pollinating insects, especially bees, and some are useful as companion plants to help control pests (see page 82). And, of course, the flowers brighten up the plot and can be used as cut flowers too.

chapter FIVE

An abundance of fruit

Fruit is an essential element of a small plot. Fruit trees and bushes not only provide you with a supply of fresh food but also bring wildlife on to the land, especially pollinating insects. You don't need much space to establish an orchard, and fruit bushes can be planted almost anywhere.

A row of fruit trees planted just inside the boundary of this half-acre plot.

In a small space it is vital to maximize the opportunities to incorporate fruit trees and bushes. My own orchard is just 5m x 5m (16' x 16'), but I have fruit trees elsewhere too: a row of apple trees trained as a cordon along the north side of the vegetable patch; some espaliers along a dividing fence. Soft fruit bushes can be popped in as single plants, planted as rows around the vegetable beds, or contained in a fruit cage.

New boundary hedgerows can be enhanced by the inclusion of species that produce edible fruit. I have seen trained apple trees used to good effect in creating a productive barrier around a small orchard in which chickens were kept. The trees hid the unsightly livestock fence while the fallen apples were soon eaten by the chickens. So, think vertical when planning the use of space, and don't waste the grass under the trees!

The orchard

Traditionally, orchards had large standard trees, but modern cultivars are much smaller and so need less space. On a commercial scale, growers plant upwards of 400 trees to the hectare (2½ acres)! Choosing your trees can be the difficult bit – with more than 700 apple cultivars alone to choose from. Think about the ways in which you want to use your fruit, and that may help you select the trees. For example, you could opt

Small spaces are fine for fruit trees, especially the 'bush' types.

By growing heritage varieties you are doing your bit to conserve the genetic diversity of fruit trees.

for a selection of cooking, dessert and cider apples, with some pears, plums, damsons and possibly cherries.

Cultivars

When choosing cultivars (cultivated varieties), think about geographical location, as cultivars suited to the cold winters and late springs of one region may not suit a warm sunny hillside in another, for example. So the best approach is to look for regional cultivars or even those that are local to your land, as these will be better suited to the conditions. If you are planning on selling your fruit or products from the fruit, it can be a good selling point to be able to say "made from local varieties of . . .". Not all local varieties will be prolific fruiters, and they may be more susceptible to disease than some of the more highly bred cultivars, but heritage varieties need to be conserved, so by growing them you are doing your bit to conserve the genetic

A heritage variety of apple, 'King of the Pippins'.

diversity of fruit trees. Different cultivars also vary in their resistance to disease, such as apple scab, and this is a particularly important consideration if you want to run the holding on organic principles.

Other factors to consider are harvest time and storage. Different cultivars fruit at different times, which gives you the opportunity to spread your harvest. But if, say, you are planning to juice your apples or make cider, you might decide to opt for a short harvest period so that all the apples are ready to be harvested at the same time. Some apples store well; others do not – so try to get a mix, as there is nothing worse than having a number of trees that produce great-tasting fruit, none of which will keep!

Another consideration is biennial cropping. This term refers to the habit of some cultivars to produce a bumper crop every other year, but not much in between. This is common in many of the cider apple varieties.

Most apple and pear varieties produce fruit on spurs (side shoots) rather than on the tips of their branches. This is particularly relevant if you are intending to train your fruit trees (see page 103), as you will need to avoid the few that are tip-bearing varieties.

Pollination

It is also important to think about pollination: fruit trees need to be pollinated, by bees and other insects, and most are self-sterile, which means that their flowers cannot be pollinated by the same tree or by another tree of the same variety. Instead, they need to be cross-pollinated by another variety of the same fruit, so you need to make sure that a suitable pollination partner is growing nearby. Cultivars are grouped according to when they come into flower during spring: Group 1 being very early, while Group 7 is very late in spring. Each group flowers for about 2 weeks, overlapping briefly with the group in front and behind. For example, a flowering Group 2 cultivar will probably come into flower in mid season, so could be pollinated by trees in Groups 1, 2 and 3.

Some cultivars are triploid, which means that their pollen is sterile, so they cannot be used to pollinate other trees, but they themselves require pollination to set fruit. Well-known triploids include some of the Bramley apples and the Blenheim Orange, Ribston Pippin and Tom Putt. Often you need two pollinating cultivars alongside the triploid – to pollinate the triploid,

Check the pollination group of your chosen trees to make sure that they will all get pollinated.

and to pollinate each other. But all is not lost if you only have a few trees, as pollen is carried some distance by insects, so in an area where there are lots of fruit trees cross-pollination can generally be achieved.

Location

Think carefully about the position of your orchard within the holding. Fruit trees don't like shade, preferring full sun, although some can tolerate partial shade, and they grow best on a free-draining soil with compost to boost the nutrient content and water-holding capacity. Avoid ground that dries out quickly in summer, floods in winter or has standing water, and don't forget that prevailing winds can inhibit growth considerably, so the sheltered side of a wall or hedge can be ideal – but be aware that frost pockets could occur in sheltered spots.

Rootstocks

A fruit tree usually consists of two parts: the scion, which is the above-ground part, and the rootstock, which is the bit below ground. The two parts are grafted together (see page 102). If you look carefully, you will see this joint as a slight bulge in the trunk, which should lie a short distance above the ground. The scion is taken as a cutting from the desired cultivar (because most fruit trees are cross-pollinated, they do not grow true from seed), but the root-stock influences certain characteristics of the

In this orchard there is a mix of fruit tree forms: a standard in the background, a half standard in the mid foreground, while at the front are semi-dwarf trees, reaching up to 3m (10').

Characteristics of different rootstocks

Fruit	Rootstock	Characteristics
Apple	M25	Suitable for standard; very vigorous, reaching 6-10m (20-33').
	MM111	Vigorous. As for M25 but better suited to half standards. No staking required.
	MM106	Ideal for cordon and half standard, staking required on exposed sites.
	M26	Semi-dwarf, to 3m (10').
	M9	Dwarf, very productive but poor anchorage so needs permanent staking for support. Ideal for cordons.
	M27	Very dwarf, final height just 1.5m (5'); ideal for pots and stepovers.
Pear	Quince A	Suitable for half standard; up to 4.5m (14'6")
	Quince C	Suitable for cordon; produces the smallest pear trees. Height after 5-10 years is up to 4m (13').
Plum or gage	St Julien A	Semi-vigorous, up to 4m (13'), for all plums, gages, peaches and apricots.
	Pixy	Dwarf, to 3.5m (11'6").
	Brompton	Vigorous, to 6m (20').
Cherry	F12/1	Vigorous, to 4m (13') or more.
	Colt	Semi-dwarf, productive, to 3.6m (11'10")
	Gisela 5	Dwarf, to 3m (10').

tree, such as how early in its life it fruits, its susceptibility to certain diseases (such as fireblight), and its tolerance of extreme cold – but especially its size.

Various rootstocks are available, and they produce trees ranging considerably in size. Fruit trees may be trained into a number of different forms, from full standard (the tallest) through half standard or bush to cordon or espalier (see page 106), so in addition to choosing the right cultivar you need to pick a suitable rootstock for the sort of tree you have in mind. Dwarfing or semi-dwarfing rootstocks are also available, which produce the smallest trees.

Buying and planting new trees

Most fruit trees are purchased as one- or two-year-old bare-rooted trees, ready to plant straight into the ground in winter. Bare-root is the most economical way to buy trees, and it usually has a greater rate of success, as the plants get away quickly in spring. You can buy larger, container-grown trees, but these are more expensive and can be harder to establish, as they are more prone to drought in the first years. This is because they tend to be substantial plants with a large shoot area so have a great demand for water in summer, which may not be met by the relatively small rootball.

Grafting your own trees

Fruit trees are not cheap to buy, even young ones. If you are planning on planting a number of fruit trees or you intend to propagate material from a heritage orchard, you may want to consider grafting your own. Grafting involves attaching a scion (a cutting) from a donor tree to a rootstock, so that the two parts join together to form one plant. The rootstocks that you use for grafting are much cheaper to buy than young trees, although it does take a bit of practice to perfect the technique, so expect a few failures.

There are several different ways to graft, but the basic method is the same: the scion is lined up with the rootstock and the join sealed. It's important to use a sharp knife to get a clean edge and to try to line up the cambium (the actively dividing tissue in the stem of the plant) in the scion and the rootstock. Once they are in close contact and held tight, the cut surfaces form a join. It is important that the cutting is dormant, but the rootstock can be either active or dormant. 'Splice', 'whip-and-tongue' and 'saddle' are all types of graft where the cutting and rootstock are of the same diameter. When the rootstock is larger in diameter than the cutting, a 'side', 'cleft' or 'wedge' graft is used. The diagram here illustrates a whip-and-tongue graft.

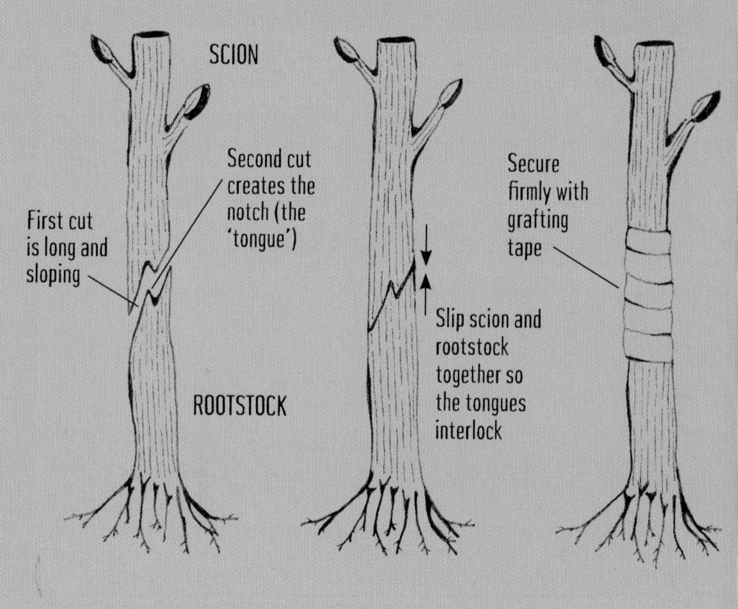

Planting of bare-rooted stock takes place in late autumn and winter, when the trees are dormant, whereas container plants can be planted at any time. Dig a large hole and place the soil in a wheelbarrow so you can mix in some good compost. For a bare-root tree, knock in your stake just off-centre and on the side of the prevailing wind. Place the tree in the centre of the hole, checking that it will be at the same planting height as it was before it was uprooted (look for the old soil mark on its trunk). Also, make sure that the graft attachment point is above the ground. Then shovel back the soil-and-compost mix, firming with your boots but not so much as to damage the roots. Strap the tree to the stake. Some smaller trees on MM111 rootstocks may not need staking. For a container-grown tree, follow the same approach but knock in the stake

Regular pruning is essential, or you will end up with a big job that takes hours and will also set back the tree.

at a 45-degree angle just outside the edge of the potted soil, so as not to damage the roots.

Make sure the tree is well watered during the first summer – give it a good soaking once a week – and keep the area around it clear of weeds by mulching. Don't let the trees fruit for a few years, so remove any blossom in spring.

Mycorrhizal fungi

Whether you buy bare-rooted or container-grown trees, it is also beneficial to use a mycorrhizal fungal powder or gel to help it establish. Mycorrhizae are symbiotic associations between certain specialist fungi and the roots of plants: the fungus colonizes the plant roots and forms what can be considered as a second root system, extending the plant's own root network and enabling it to take up nutrients from a much greater area of soil. In return, the plant provides the fungus with sugars. Mycorrhizal fungi are found in soils naturally, but in poor soils there may be fewer present, so adding a mix of mycorrhizal fungal spores to the planting hole ensures a good start. The young plants grow a denser root system that leads to strong growth and less susceptibility to drought.

Pruning

Regular pruning is essential. If you leave it several years, you end up with a big job that takes hours and also necessitates a hard prune, which will set back the tree. If you do a small amount each year, the impact on the yield will be minimal. The aim of pruning is to develop an open shape with good air circulation, which reduces the risk of disease and lets more light in to get a good crop yield. The subject of how to prune is covered in many books, but if you are not confident, go on a course or ask someone with experience to come in and demonstrate how to do it.

Most pruning takes place in winter, although you may need to do summer pruning on vigorous trees and on stone fruits (apricots, peaches, plums) to lessen the risk of spreading the fungal disease silver leaf. You will need sharp secateurs and loppers or a pruning saw. First, remove any dead, diseased or dying branches. Shorten recent growth on each main branch by about a third, to stimulate the formation of new branches and spurs. Don't touch the new side shoots, as these will develop new fruit buds (assuming the tree is spur-bearing – see page 98), unless they are crowding out other shoots or crossing or growing towards the centre of the tree.

Training fruit trees

Fruit trees can be trained to grow into a certain shape, such as a fan. The main reason to train a fruit tree is to take best advantage of the available space, for example against a sunny wall. It is also thought to improve the fruit, leading to fewer but larger and better-tasting fruit, as the

In this vegetable garden, fruit cordons have been trained to form a decorative arch.

trees put all their energy into fruiting rather than into producing more leaves.

Among the most popular trained forms are cordons, fans and espaliers, which are ideally suited to growing against a wall or to creating a vertical barrier on the plot, using a trellis or similar support system. With these forms of training, the trees get more light and better airflow, which helps the fruit to ripen and reduces the incidence of disease such as mildews. Growing fruit trees against a wall also allows you to take advantage of the microclimate and to grow varieties that need more warmth, such as apricots and peaches, while a trained tree is also easier to protect from frost and birds.

Cordons

When space is in short supply, a cordon-trained tree can enable you to produce a relatively sizeable harvest. A cordon is a tree that has a single stem with lots of short, stubby side shoots which bear fruit. This form is suitable for spur-bearing apples and pears, since these fruit on short side shoots, and the yield can reach as much as 10kg (22lb) per cordon. Tip-bearing varieties are no use, as the tips will be pruned, so you would never get any fruit. Cordons are very useful as internal dividers on the plot.

Typically, cordons are grown in a row, each tree planted at an angle of 45 degrees, but there are variations, such as the 'U', with two vertical

Cordons need a strong framework of supporting wires, fixed to a fence or wall or to sturdy stakes.

cordons, fans and espaliers are ideally suited to growing against a wall or to creating a vertical barrier.

branches, or even versions with three or four verticals per tree. It's important not to use vigorous rootstocks, as they are difficult to control. If you grow a mix of cultivars, you shouldn't have to worry about pollination, and can also get a useful combination of cooking and dessert fruit.

A cordon is usually trained at an angle of about 45 degrees, with close planting of 70cm (28") between the plants in the row. Support is essential, so a cordon is planted against a wall or fence or supported with posts about 3m (10') apart, with a framework of three or four horizontal wires about 50-60cm (20-24") apart.

When planting, ensure not only that the grafting join is above the ground but also that the wood above the graft is uppermost and that the line of the join points towards the ground, to reduce the risk of the trunk breaking. Tie long bamboo canes to the wires at a 45-degree angle to support the young trees.

Suitable rootstocks for cordons are M27, M9 and M26, depending on the soil type, and Quince C. On poorer soils it can be better to choose slightly more vigorous rootstocks. If you have multiple rows of cordons, leave a space of around 60-70cm (24-28") between rows. Once planted, cut back any long side shoots to three buds and leave the leader (main) shoot unpruned. Cordons should be pruned in late summer by cutting back any new shoots that exceed 20cm (8") or three leaves above the basal cluster of leaves (the cluster around the base of the current year's growth). Leave any shorter new shoots. New shoots on existing side shoots are pruned to one leaf above the basal cluster, and any upright growth is removed. Allow the cordons to reach an eventual length of 2m (6'6") but no more, so harvesting is easy.

Fans, espaliers and stepovers

These differ from cordons in that the trees have a number of branches. In the case of an espalier, the branches extend out sideways from the main stem, while in a fan the branches radiate out. The best apple rootstock is MM106, or M111 on poorer soils, and for pears Quince A. Use St Julien A for plums, apricots and peaches. Support is essential for fans and espaliers, so they need a framework of horizontal wires spaced at 40cm (16"). Allow 4-6m (13-20') between plants.

Any fruit tree can be trained into a fan. A one-year-old grafted tree (known as a maiden) is cut to 60cm (24") in autumn, and the following spring four to six buds are allowed to grow as branches. These are trained by attaching them to the wire framework to create radiating arms.

Espaliers are also produced from maiden trees. The tree is cut back to a bud that lies about 15cm (6") above the lowest wire. This bud is allowed to become the vertical branch, and the shoots from the next two buds below are trained, one to each side, to create the first tier of branches. This is achieved slowly, by gradually tying the shoots down so that they are eventually lying horizontal and are tied to the wire (but not too tightly). This process is repeated in subsequent winters, so that several tiers of shoots are produced.

A stepover is simply an espalier with just one tier, and the main leader tip removed. Like cordons, they are useful for growing around large vegetable beds and dividing up areas of the plot.

It is important to prune these trained trees regularly to maintain their shape. In late summer, fans and espaliers should be pruned by cutting back new side shoots to three leaves from the basal cluster, while shoots on existing side shoots can be cut to one leaf above the basal cluster. Any suckers from below the graft should be removed.

Soft fruit

Soft fruit is expensive to buy, so it is an obvious choice for the small plot. With careful planning you can harvest soft fruit for 8 months of the year, starting with early strawberries under cover in early spring and ending with autumn raspberries, which can be harvested until the last frosts. For the remaining months you could rely on soft fruit in the freezer.

A row of fruit trees trained as espaliers can create an attractive and productive boundary.

With good planning, you can be harvesting soft fruit for 8 months of the year, from early spring to the last frosts.

Fruit cages

Soft fruit tends to be grown in blocks of space, simply because the juicy fruit needs protection from birds and the easiest way to achieve that is to build a fruit cage. You can grow them in rows where space allows, and cover them with netting, but expect some losses.

Commercial fruit cages have a tubular frame with polypropylene mesh over the sides and roof. Most of the modular systems that are available are based on 2m (6'6") lengths. However, a fruit cage is relatively easy to construct should you wish to save money and build your own. Also, you only need a tall walk-in one if you are growing raspberries; a lower one would

The netting on a fruit cage can be damaged in winter if there are high winds or heavy snow, so often it is taken down in autumn.

suffice for currant bushes and strawberries, although it's not so easy to manage.

Ideally, a fruit cage should be positioned on a level site in full sun or in sun and partial shade, and avoiding a site that might catch any late frosts. A space as small as 4m x 4m (13' x 13') is enough to accommodate a number of bush plants such as blackcurrant, redcurrant and whitecurrant bushes, jostaberries, gooseberries and several rows of raspberries and blackberries, as well as strawberries and rhubarb. Don't forget to leave space to move around between the rows of plants once they are mature! The cage pictured above is 7m x 5m (23' x 16').

Soft fruit are permanent crops, so it pays to clear the weeds thoroughly and prepare the soil well. The soil needs to be free draining with good fertility. Either dig over the site to remove any compaction and perennial weeds and mix compost into it, or, if you are using the no-dig method, cut any vegetation close to the ground, ease out any large perennial weeds and cover with cardboard and a thick layer of compost. You can plant directly into this.

To reduce the growth of future weeds you could cover the ground with bark chippings, but be aware that they take up nitrogen from the soil as they rot down (see Chapter 3, page 50). I prefer to cover my no-dig plots with ground-cover fabric to suppress any weeds, as well as to conserve water. If you don't use a ground cover, take care not to hoe too close to the bushes, as soft fruit bushes have relatively shallow roots that can be damaged easily.

Soft fruit to choose from

The table on the following pages lists some of the soft fruit that you might consider growing on your plot. Choose thornless varieties where available, for easy harvesting and maintenance.

Poultry in the fruit cage

Don't forget to maximize the use of the ground in the fruit cage. If you have grassy areas between the rows of plants, you could use the space for birds. Chickens are the obvious candidates, but it's not wise to have them in the cage from the time the fruit buds are forming until leaf drop. But the winter months, when the plants are dormant, are perfect, as the chickens cannot do any damage and will clear the ground of insects and other pests. They are particularly good at clearing away overwintering sawfly pupae, which devastate gooseberry plants. The chickens will scratch, unless you have feathered legged breeds such as Brahmas, but this could help to keep the beds weed-free. Watch out for them creating deeper dust-bath areas, which would damage the shallow roots of the fruit bushes.

Ducks tend not to damage the fruit and don't scratch, so could be housed in the fruit cage for much of the year, depending on the size of the cage and the number of birds. Ducks will tackle pests too, especially snails and slugs and their eggs, and they could be introduced into the fruit cage when the sawfly larvae are active, in late spring through to mid summer.

Smaller areas can be used for quail, which could be kept in a house and run within the fruit cage, giving double protection from the fox. Be careful, however, if you plan to let quail run free-range in the cage, as they have a habit of flying up when scared and bashing into the netting.

Fruit-cage netting is not fox- or badger-proof, so remember to put an electric line around the bottom if the cage is accessible to predators. See Chapter 7 for more about chickens, ducks and quail.

Soft-fruit planting guide

Fruit	Planting guidance	Cultivation notes
Blackberry	Allow 2-4m (6'6"-13') between plants. Grow in sunny position for sweetest fruit. Support with horizontal wires or fencing, or plant against a wall. Can be trained up a trellis or arch.	Mulch in spring. Tie in shoots as they grow. Once fruit is harvested, prune by cutting old shoots to the ground. Suckers readily when tips touch the ground. Modern hybrid varieties tend to have spineless canes and produce larger fruits.
Blackcurrant	1-1.3m (3'3"- 4'3") between plants. Can be grown as a cordon. Plant deeper than the original depth to encourage new shoots.	Start pruning in its second year, once plant has eight good shoots: remove a quarter of the oldest stems in winter, cutting to the base and creating an open cup shape for good airflow. Blackcurrants benefit from a thick mulch. Feed in summer with comfrey tea.
Blueberry	Allow 1m (3'3") between plants. These plants require an acidic soil of about pH 5.5 or lower, so unless you have a naturally acidic soil, you will need to raise the plants in pots with ericaceous compost, or apply plenty of acidic organic matter, such as pine needles, to their bed to reduce the pH. Use two different varieties to ensure cross-pollination.	Prune new shoots to 60cm (24") in summer, and winter-prune to remove dead or weak shoots. Mature plants should have a quarter of the oldest stems removed each year.
Gooseberry	Allow 1-1.3m (3'3"-4'3") between plants. Can be grown as a cordon or as a lollipop standard (shoots trained into a ball shape with a long bare stem).	Once planted, cut back main branches by half and remove the lowest shoots below 20cm (8") from the ground to give access for weeding under the plant and allow airflow, which reduces the risk of mildew. Gooseberries fruit on old wood, i.e. branches from the previous year or before, which produces fruiting spurs.
Honeyberry (*Lonicera caerulea*)	Allow 1.5m (5') between plants. Plant in pairs or groups to aid pollination. Prefers full sun and is drought tolerant. Support with horizontal wires or fencing, or plant against a wall.	An edible honeysuckle that grows to 1m (approx. 3') tall, producing blue fruit in late spring. Flowers in late winter, so hand pollination may be needed. Once established, prune in summer after harvesting, thinning overcrowded and weak shoots to the base. Remove tips of young shoots to encourage flowering laterals.
Jostaberry	Allow 2m (6'6") between plants. Grows to 2m (6'6") in height, and can be grown as a shrub or trained against a wall.	A blackcurrant–gooseberry cross. Fruit forms on the one-year-old or older canes. Prune in winter by removing about half of the new growth. Mulch well.

Fruit	Planting guidance	Cultivation notes
Juneberry / saskatoon (*Amelanchier alnifolia*)	Allow 1.5m (5') between plants.	A North American fruit, similar to blueberry in taste and shape. Also grown as an ornamental. Self-fertile. Bushes can grow to more than 4m (13') if unpruned. Prune to remove any low, spreading branches to encourage airflow, and cut back the shoots to 2m (6'6") in length.
Kiwi fruit	Allow 3-4m (10-13') per plant. Plant against a wall or fence, or train over a trellis or arch. Prefers a slightly acidic soil.	Kiwi fruits are vigorous plants, extending up to 9m (30'). Plants may be either all-female, all-male or self-fertile. If self-fertile, you can plant just one; all-female plants need a male plant to pollinate them in a ratio of 3 to 4 females per male. Mulch well so soil doesn't dry out in summer. Tie in the long stems as they grow. Once the plants have established, prune the side shoots in summer and winter. Each winter, cut one-third of the oldest side shoots back to a bud about 5cm (2") from the main stem.
Lingonberry / cowberry	Allow 0.5m (20") between plants. Prefers moist, acidic soils.	A low-growing evergreen bush to 20cm (8"), with attractive red berries that are bitter and need cooking. Very cold hardy. Could be planted as a ground cover along hedges.
Loganberry	Allow 2m (6'6") between plants. Canes extend to about 2m (6'). Prefers full sun. Train along a fence or wall, or provide supporting wires.	A cross between a blackberry and a raspberry. Fruit forms on one-year-old canes. After fruiting, cut canes to the ground.
Raspberry	Plant canes 40-50cm (16-20") apart, with a row spacing of 1.5-2m (5'-6'6") for summer raspberries and 2m (6'6") for autumn raspberries.	Mulch in spring. Summer-fruiting canes need supporting on horizontal wires supported by posts unless grown against a sunny wall or fence. They fruit on the last season's growth, so prune once the fruit are picked by cutting old brown canes to the ground, and tie in the new shoots. Autumn-fruiting varieties don't always need supporting. They fruit on the current season's growth. Cut the canes to the ground in late winter or early spring. Or, for double cropping, leave unpruned, and a modest early crop will form on the old canes. After the first crop cut the old canes to the ground, and fruit will form on the new canes.
Redcurrant	Grow as bushes, cordons or espaliers. Allow 1-1.3m (3'3"- 4'3") between bushes.	Treat more like gooseberries than like blackcurrants. Once established, in winter prune main branches by one-third, just above an outward-pointing bud, trying to keep the centre of the plant open.

Fruit	Planting guidance	Cultivation notes
Strawberry	Plant in open sunny position in rich soil, 45cm (18") apart, in rows 75cm (30") apart. Strawberries do better planted through ground-cover fabric, which reduces weeding and retains moisture.	Benefits from regular feeding throughout the growing season. After fruiting, cut back the foliage and allow just the strongest runners to root to provide replacement plants; remove the rest. Replace plants every 3 to 4 years.
Tayberry	Allow 2.4m (8') between plants, as they are vigorous, with canes reaching 2.5m (8') in length. Grow against a sunny wall or use supporting wires. The plants are very heavy when fruiting.	This is a cross between a blackberry and raspberry and is a prolific cropper. 'Floricane' varieties fruit on the canes from the previous year: cut to ground level after harvesting and tie in the new canes. 'Primocanes' flower and fruit on the current year's growth and should be cut down after fruiting.
Whitecurrant	Grow as bushes, cordons or espaliers. Allow 1-1.3m (3'3"- 4'3") between bushes.	Treat more like gooseberries than like blackcurrants. Once established, prune main branches by about one-third in winter, just above an outward-pointing bud, trying to keep the centre of the plant open.

Soft-fruit hedges or screens

Don't forget that fruit such as currants and gooseberries can be grown as cordons and fans too (see pages 104-6), using supporting wires and posts or trained against a wall to take up minimal space. Choose varieties that have a vigorous and upright growth habit. Gooseberry cordons should be planted about 35cm (14") apart, while currant cordons need a bit more space, at 40cm (16"). The rooted cuttings are planted at a 45-degree angle and the leading shoot is trained to a supporting bamboo cane which is secured to the wires. Remove any side shoots less than 15cm (6") from the ground. The following year, prune the shoots twice: first in mid summer, when all the side shoots should be cut to five leaves, and again in winter, when the side shoots are taken back to two buds. Prune the growing tip by one-third. Allow the cordon to reach about 1.8m (6') in length, although longer lengths are possible on a wall where more support is available.

If you don't have room for a soft-fruit bed or cage, then a fruiting hedge made up of blackcurrants, gooseberries, redcurrants and whitecurrants is a good alternative. Plant at about two plants per metre (3'), and either allow them to grow into bushes and prune to about 1m (3') in height, or train them as cordons, using stakes and wire to support the shoots. Unsightly trellises or chain-link fencing can be disguised with climbing blackberry, tayberry and loganberry plants.

Taking cuttings

It's easy to propagate soft fruit and produce a large number of plants for the plot very cheaply.

In this orchard, a cordon of gooseberries has been planted between two rows of fruit trees to maximize use of the land.

If you don't have much room for soft fruit, one option is a fruiting hedge, grown as bushes or trained cordons.

Currants, gooseberries and jostaberries are best propagated from hardwood cuttings taken in winter, when the plants are dormant. Cut off a healthy stem at the base that is at least 30cm (12") long and as thick as a pencil with about six buds along its length. Remove the soft growth at the tip. Then trim it to about 25cm (10"), cutting below a bud at the base and just above a bud at the top. Make a deep slot in your nursery bed, pour some sharp sand into the base of the slot, and insert the cuttings so that half of their length is in the ground, leaving 20cm (8") between cuttings. Firm the soil with your feet and leave the cuttings in place until the following autumn, when they can be transplanted to their final place. If you don't have a nursery bed, you could use a large container.

Blackberries can be propagated by stem cuttings too. After harvesting the fruit, simply take a leafy length of cane of about 15cm (6") and insert it into the nursery bed to a depth of about 8cm (3"). Water well and keep moist. They should produce roots within a month. Leave them in place and wait until the following autumn before transplanting.

Edible boundaries and barriers

On a small plot it is essential to make use of all of the available space, and that includes the boundaries. You may have a site with an existing hedge boundary, or a wall or fence that could be used for food production. If not, new hedges can easily be established around the boundaries and within the holding to subdivide spaces.

Hedges will not only look attractive all year round but also provide a supply of produce – fruit, nuts, hips, edible flowers, leaves and spices. They are excellent habitats for wildlife, attracting birds and many beneficial insects. An edible hedge can be a large, full-sized boundary structure or just a single line of shrubs of the same species.

A mixed hedge will provide a variety of fruit and nuts, but is more difficult to prune, as each species has its own requirements. Also, the selection of species would need to be considered carefully, as the inclusion of faster-growing, more aggressive species, such as elder and blackthorn, can suppress the growth of slower-growing species, such as hazel. The management of a single-species hedge is more straightforward and may therefore be preferable. Species that would suit management as a hedge include pears, wild pear, elder and cherry plum.

Edible hedges are most productive if allowed to grow to their natural height, which may be over 4m (13') tall – so if you want a hedge that is barely a metre (3') or so in height then you would be better to opt for bush species rather than trees, for example a bush cherry rather than a tree, as a tree will not flourish if pruned hard every year.

Hedgerow species

A traditional selection of species for an edible hedge in northern Europe includes hazel, blackthorn, hawthorn, crab apple, dog rose and blackberry, but there are other options too (see table, opposite). These plants will grow quite quickly and provide a good yield of fruit and nuts for making hedgerow preserves and other products. They will also create a stock-friendly barrier for livestock around the plot. Don't keep the hedge looking neat and tidy, as it needs to be able to grow to produce a decent harvest.

If you have an existing blackthorn or hawthorn hedge along the edge of your plot, you can convert it to a more varied edible hedge by thinning it and planting edible shrubs and trees in the gaps and alongside it. If the existing hedge is a single row of plants, you can widen it to two or three staggered rows to boost productivity.

Planting your hedge

Hedge planting is usually done during winter, when the plants are dormant. Saplings can be purchased from hedging specialists, which sell bare-rooted stock ready for immediate planting.

Make sure you have prepared the ground before the plants arrive. Remove as much vegetation as possible along the planned line of the hedge, ensuring that perennial weeds are removed, and enrich the soil by covering it with a thick layer of compost or well-rotted manure. To save on labour, you can clear the ground by covering it with sheeting or matting several months before the date of planting. The saplings will be slit-planted, so make sure you can push your spade into the ground – if not, loosen the ground with a fork and remove any stones.

Edible trees and shrubs for hedges and barriers

Common name(s)	Latin name	Description
Autumn olive	*Elaeagnus umbellata*	Vigorous shrub or small tree, growing to 5m (16'), with silver-green leaves, tiny yellow flowers and small, dark red, tasty fruit. Nitrogen fixer, so will boost neighbouring plants.
Barberry	*Berberis vulgaris*	Deciduous shrub growing to 3m (10'), with red edible fruits.
Blackberry	*Rubus fruticosus*	Vigorous and spreading climber. There are a number of hybrids with spineless canes, which bear larger fruit.
Blackthorn	*Prunus spinosa*	Fast-growing, prickly shrub creating dense cover. Grows to 4m (13'). White flowers early in spring; sloes in autumn.
Cherry plum	*Prunus cerasifera* Myrobalan Group	Traditionally grown as a shelter belt for orchards, a row makes a good windbreak 6-8m (20-26') high. Yellow or red plum-sized fruit in late summer, which are sweet and juicy.
Chokeberry, black	*Aronia melanocarpa*	Grows to 1.5m (5'), with glossy black berries in autumn. Use as a currant.
Cornelian cherry	*Cornus mas*	Large shrub or small tree, growing to 4m (13') or more but can be pruned. Bears cherry-like fruit with a plum flavour.
Crab apple	*Malus sylvestris*	Usually allowed to grow as a standard tree in a hedge, reaching 8-9m (26-30'). White flowers in late spring and crab apples in autumn, which are ideal for making jellies and wine.
Dog rose	*Rosa canina*	Fast-growing climber with arching branches. Pink-white flowers in early summer and edible red hips in autumn.
Ebbinge's silverberry	*Elaeagnus* x *ebbingei*	Hybrid shrub, evergreen for year-round cover, grows to 5m (16') but can be pruned to desired height. Scented flowers are produced in autumn followed in late spring by small red fruit that ripen to become rich in flavour.
Edible rowan	*Sorbus aucuparia* var. *edulis*	Attractive tree that can grow to 15m (49'), with edible fruit, which are larger than those of the wild rowan. Can be used in preserves.
Elder	*Sambucus nigra*	Fast-growing shrub that can reach 8m (26') and outcompete neighbouring plants. Can be pruned. Flat flower heads of edible small white flowers in spring, black fruit in late summer.
Flowering quinces	*Chaenomeles* spp.	Bush species that can be grown as a single-species hedge or in a mixed hedge. Can reach 4m (13') or more, but may be pruned to height. Bears apple-shaped fruit that can be used in much the same way as the true quince.
Hawthorn / quickthorn	*Crataegus monogyna*	Fast-growing, thorny and hardy shrub or small tree, reaching 6m (20'), with white flowers in late spring and edible red haws in autumn that can be used in preserves.

Common name(s)	Latin name	Description
Hazel	*Corylus avellana*	Shrub or small tree, reaching 8m (26') but can be pruned to reduce height. Produces nuts in late summer and autumn but needs cross-pollination for good harvest. Can be coppiced to provide poles and fuel.
Raspberry	*Rubus idaeus*	Will carry fruit if planted in a sunny position or alongside a hedge.
Rugosa rose	*Rosa rugosa*	Fast-growing bush-forming species up to 2m (6'6"); attracts pollinators. The edible flowers can be harvested as well as the large hips produced in autumn.
Salmonberry	*Rubus spectabilis*	Forms a decorative hedge to about 1.5m (5') high. White flowers are followed by tasty orange fruit that look like raspberries.
Sea buckthorn	*Hippophae rhamnoides*	Large nitrogen-fixing shrub, growing to 6m (20'). Prefers a sunny site with well-drained soil. Both male and female plants needed for fruit. Fruit is tart, but high in vitamin C.
Service tree	*Sorbus domestica*	Small-to-medium-sized tree, reaching as high as 25m (82'), with red plum-sized fruit, which need to be frozen to make them edible.
Snowy mespilus	*Amelanchier lamarckii*	Shrub or small tree, reaching 12m (40'), with white flowers. The red/purple currant-like berries are apple-flavoured. Attractive orange-red foliage in autumn.
Szechuan pepper	*Zanthoxylum schinifolium*	Aromatic shrub that grows to 4m (13'), ideal for creating a single-species hedge. The dark red peppercorn-like fruits are used as a spice.
Wild pear	*Pyrus communis*	Small tree growing to 8m (26'), with white flowers. The fruit can be cooked or used for perry (pear equivalent of cider).

Keeping the weeds at bay will be critical to the successful establishment of the hedge, so you can either cover the ground with a 1m (3')-wide length of ground-cover fabric, and plant through this, or surround the saplings with a thick layer of mulch to suppress weeds. The recommended spacing is 40-50cm (16-20") – this is generous, as you want the plants to bear lots of fruit and nuts. If planted more closely together, they would create a tight, secure hedge, but the yield would be considerably lower. For a wider, more productive hedge, plant two staggered rows at 50cm (20") distance. If your soil quality is poor, you can boost the chances of the plants establishing well by using mycorrhizal fungi powder, which is easily applied by dipping the bare roots in the powder before planting (see page 103).

If you are using ground-cover fabric, first secure the edges by pushing them into the ground with a spade. To plant your sapling, make a slit in the membrane to the width of your spade. Then push in your spade and wiggle it to make a slit in the soil and insert your sapling, making sure that it is planted no more deeply than it was before it was uprooted (look for the old soil mark on its stem). Carefully remove the spade and firm the sapling into place with your boot. If you have rabbit problems, protect the saplings using a plastic spiral guard and bamboo cane. Once in place, trim back your saplings to about 30cm (12") so that the plants form lots of low branches, filling in the hedge.

The plants need to be pruned regularly, but don't prune the whole hedge at once. Most hedgerow plants produce their fruit on new wood, so it's better to prune one side of the hedge one year and the other side the next year, so you always have some branches fruiting. Keep an eye on the more vigorous ramblers, such as blackberries, as they will need to be cut back regularly to prevent them suppressing the growth of the other plants.

Barriers can also be created within the plot using larger vegetable species. For example, a row of asparagus, globe artichokes, Jerusalem artichokes, sweetcorn or even sunflowers will create a barrier within a few months.

A newly planted edible hedgerow on our plot, which was prepared with a thick layer of mulch and covered with a membrane to keep down the weeds.

chapter six

Tree crops & forest gardens

You may think the only sort of trees you have space for on your plot are fruit trees, but with careful planning it's remarkable how many trees you can fit in to a small area. You can even grow wood for fuel, if you have a wood burner in your home.

Willow grown for fuel.

When considering where to include trees on a small plot, you need to think about how to make best use of the space. One option is to use the space in more than one way, to get the most from the land, for example by running your chickens under a willow crop. Forest gardens – in which edible trees, shrubs and low-growing plants are combined in a natural woodland-like manner – make good use of vertical space. They can be surprisingly productive, and are not as labour-intensive as vegetable plots.

Fuel supply

The modern hybrid willows have been selected and bred for their fast growth. Some can grow 2-3m (6'6"-10') in their first year, reaching a maximum of about 10m (33') if allowed to mature. A 15m x 50m (49' x 164') strip of fast-growing willows could produce a ton of wood a year. This may not be enough to heat a whole house, but it is more than sufficient for a small wood burner that is used in the evenings and at weekends. There are many different varieties of hybrid willow, varying in their rate of growth and tolerance of different conditions as well as their bark colour.

Coppicing

Willows are usually grown for about 5 years and then the stems are cut to the ground, in a process called coppicing. The shoots grow back and can be harvested again and again. Coppicing is a traditional system of woodland management used for tree species such as hazel and sycamore, which are harvested for stakes, poles, woodchips and firewood. The woodland is divided into sectors, the number of sectors being the equivalent to the number of years that the trees will be allowed to grow: for example, hazel and sycamore tend to be grown on long rotations of 8 to 15 years. So for a 15-year rotation the woodland is divided into 15 sectors, and one sector is harvested each year. This not only provides a sustainable, annual supply of wood products but also provides a range of different habitats for wildlife, from open woodland in the newly coppiced sectors to dense woodland ready to be harvested. This form of management supports the greatest diversity of plants and animals.

Growing poplars

Another option for fuelwood is poplar. Hybrid poplars can grow several metres (10 feet or more) in height a year; the fastest-growing reaching 15m (49') in just 5 years. Like willow, they can be established from cuttings. They are usually left longer than willow before harvesting, but they yield good-quality logs. We have poplar growing around our small plots of willow, forming a boundary, and we harvest them every 8-10 years, but they could be harvested every 5 years. A recent harvest of 15 trees yielded one ton of logs. We pollard the poplars (see page 122), so the new shoots grow above deer height.

Mature poplars ready for harvest.

Recently pollarded poplars: the new shoots already 2m (6'6") long in just 5 months. The willow visible behind is ready to be harvested.

Coppiced woodland provides a sustainable supply of wood products and also a range of habitats for wildlife.

A similar management system is used for willow, with a short rotation of about 5 years. To create the rotation, divide your 15m x 50m (49' x 164') strip into five sections of 10m x 15m (33' x 49'), and harvest one section each year.

To establish your willow coppice you need a lot of willow cuttings. For example, a 50m x 15m strip will need 500 plants. These can be bought from specialist nurseries in winter. The cuttings are simply 30cm (12") lengths of willow supplied in bundles of 50 to 100, which will root readily once pushed into the ground. Just make a 20cm (8")-deep hole in the ground with a metal bar or sturdy rod, insert the cutting, and firm it into place.

Coppice willow is planted in rows, with a spacing of 1m (approx. 3') between plants and 1.5m (5') between rows. You can grow them closer, but

A nut orchard or nuttery is a versatile addition to the plot, supplying a **protein-rich food to complement fruit.**

the stems will be thinner, which may be fine if you are growing for woodchip or basket weaving but are not so suitable for fuel. It is important to control the weeds in the first few years of growth in the coppice cycle as they will compete with the young willow plants, and this is most easily achieved by using ground-cover fabric or thick mulch, which you plant through. The first winter after planting, cut the shoots almost to the ground to encourage lots of new shoots. Replace any cuttings that have died. You can start your coppice cycle the next winter by cutting the first section to the ground, although you won't be harvesting much for the first few years!

Pollarding

If you would like to run poultry under your willow coppice, to make more use of the space, it's better to pollard the trees rather than coppice them – that is, to cut them at least 60cm (2') above the ground. This creates a willow with a sturdy lower stem and a cluster of new shoots above the ground, out of reach of the birds. Once the willows are established, remove the ground cover and seed with a grass mix. Willows create a dappled shade but do not shade out grass completely, so vegetation will grow for your birds. The birds can be introduced once the vegetation cover is established.

Pollarding is also preferable if you have a problem with rabbits, as they too can damage the young tree shoots. Pollarding to 90cm (3') lifts the new shoots above rabbit height. I have seen willow successfully pollarded even higher, at 1.5m (5'), which means that the area could be used for grazing sheep or goats. If you are intending to pollard, then buy cuttings about 1m (3') long and don't cut them back at the end of the first year.

Shelter belts

Tree crops such as willow can be grown not only as a sustainable source of firewood but also as a shelter belt for young fruit trees. Wind sets back the growth of plants considerably. On our first holding, we didn't realize just how much wind damage we were experiencing. We were located on a gently rising slope and didn't consider the holding to be windy, but the fruit trees took forever to get going. We then planted a double row of willow, which grew several metres (at least 10 feet) in the first year, providing an almost instant windbreak, and the benefits were noticed almost immediately too. The willows produced long stems, which we harvested and sold to a basket maker to be used for weaving screens and edging for raised beds.

Willow cuttings in a shelter belt are planted more closely than willows planted for fuel. For example, plant two rows through ground-cover fabric, with 25cm (10") between plants and 50cm (20") between rows.

Mushroom logs

If you have a shady spot and some spare logs, then you can grow your own shiitake or oyster mushrooms. To grow them you need a supply of 1-2m (approx. 3-6') lengths of freshly cut hardwood logs, about 8-20cm (3-8") in diameter, cut in winter when there is high moisture content. You also need some fungal spawn - tissue impregnated with fungal mycelium - to inoculate your logs. The easiest form in which this comes is as plugs on hardwood dowels, but some suppliers provide impregnated sawdust.

The logs are inoculated by drilling rows of holes in the wood, 2.5cm (1") deep and 15cm (6") apart, with 10cm (4") between the rows to create a diamond pattern. The plugs or sawdust are inserted and sealed with wax, and the inoculated logs then stacked outside in dappled shade. You can expect your logs to fruit after 5 to 18 months. Since the time of fruiting is uncertain, you can force your logs into fruiting by 'shocking' them. This involves tapping the logs on the ground and immersing them in icy-cold water for 24-48 hours. Under ideal conditions, a log can be expected to fruit for up to 2 years, with two harvests a year.

A pile of inoculated logs.

The forgotten nuttery

Long forgotten in modern cultivation systems, a nut orchard or nuttery is a versatile addition to the holding, supplying a protein-rich food to complement fruit. Nut-bearing plants are easy to grow, and, as with fruit trees, there is no need to designate an area just for the nut trees, as they can be planted in hedgerows and other convenient locations.

The two best-known nut trees suited for small spaces are the cobnut (*Corylus avellana*) and filbert (*Corylus maxima*): cultivated varieties bred from the common hazel. The husk of the cobnut is shorter than the nut, while the filbert (also called the full beard) is a longer nut with a fringed husk that extends beyond the nut. Other options for nut trees include almonds, sweet chestnuts and walnuts, although these do grow into sizeable trees.

Growing cobnuts and filberts

Cobnuts, filberts or hazels should be planted at 4m (13') intervals, in staggered rows. You need to buy a mix of varieties to ensure cross-pollination, as most are self-sterile. The trees can be bought as bare-rooted stock and should be planted in autumn into prepared ground mixed with compost. Once planted, cut back the stems and, if not using ground-cover fabric, mulch well to suppress weeds. Use tree guards if there is a risk of rabbit or deer damage. If the soil is too fertile, you will get excessive leafy growth in young trees at the expense of female flowers, so you may need to snap off leafy shoots to boost flower production. It takes about 4 years before the trees bear nuts. (See also Resources.)

The nuts are harvested by hand in late summer when they are in their green state, and this can be quite time-consuming, as you need to inspect the nuts for damage and disease. If you drop them in water, diseased and damaged nuts will float to the top. Dry them in their shells until they turn brown, when they will have developed their full nutty flavour, then remove the husk. Dried nuts can be stored in a single layer, or in a nut net, in a dry ventilated place protected from mice. If you have a large harvest it may be worth investing in commercial nut crackers to extract the kernels. Under ideal conditions they can store for a year, but to ensure that they don't go rancid you can freeze the shelled nuts. (Sweet chestnuts are different – since they are mostly carbohydrates, they go hard when dried.)

Tiny round holes in the shell are a sign that you have nut weevils. These pests lay their eggs in young nuts, and when the larvae hatch they eat the developing kernels. They exit the nut by boring a small hole in the shell. The pest can be controlled by running chickens under the trees in late spring to the middle of summer, when the weevils are active on the ground.

It's not essential to prune hazels or their cultivars, as they tend to grow into large bushes, but if you want to be able to reach all the nuts you should keep the plant to about 2m (6'6") high by pruning. In winter, remove any diseased and dead wood and thin out shoots, directing them outwards to create a open bowl shape with about six to eight framework branches.

An edible forest garden

If you don't have the time or inclination to cultivate fruit and vegetables on your plot, a forest garden may be the answer. It's not a quick fix, but in time a forest garden can provide a supply of food and other useful products with minimal cultivation and maximum biodiversity. Pioneered by Robert Hart, and more recently by Martin Crawford, a forest garden is a system based on trees, shrubs and other perennial plants that mimics the structure of a young natural woodland. (See also Resources.)

There are two key elements of a forest garden: the productive plants and the perennial support plants. The productive plants include not only edible crops (fruit, nuts, edible stems, leaves and roots) but also medicinal herbs, plants that provide fibre and others that may be harvested for timber. The support plants perform roles such as providing ground cover or shade, attracting beneficial insects, fixing nitrogen or drawing up minerals from deep in the soil. The incredible richness of species in such a system

In this view of Martin Crawford's forest garden in Devon, there is a shrub layer of *Phormium* and *Elaeagnus* and a canopy layer of Italian alder and fruit trees, while the ground-cover layer includes lemon balm and comfrey.

A forest garden is like a cross between an orchard and a forest, designed to let plenty of light reach the ground.

means that there are fewer diseases and pests, partly because there are no large blocks of crops to attract them, and also because the biodiversity leads to an array of natural predators.

A layered approach

A forest garden can be likened to a cross between an orchard and a forest, rather than a mature woodland. In a tropical rainforest or a mature temperate woodland, little light penetrates the canopy, and this restricts what can grow below. But a forest garden is designed to let plenty of light to reach the ground, allowing up to seven different layers of plants to flourish:

The canopy In a forest garden the canopy trees rise to above 10m (33'), but they are deliberately few and far between. They are too tall to be useful for harvesting edible crops, so they are usually timber trees, such as oak or ash, or trees that can fix nitrogen, such as the black locust (*Robinia pseudoacacia*) or walnut for nuts. However, in a small forest garden, space constraints mean that this layer is usually omitted, or there may be just one or two tall trees.

The lower canopy At 4-9m (13-30') high, this is the main tree layer in a small forest garden, comprising species such as Italian alder (*Alnus cordata*), wild service (*Sorbus domestica*), apple, pear, medlar, quince, mulberry, fig, hazel and willow. The trees are spaced well apart.

The shrub layer At 1-3m (3-10'), a variety of woody shrubs can be grown under the lower-canopy trees, such as currants, gooseberries, hawthorn, blackthorn, *Elaeagnus* species, flowering quinces (*Chaenomeles* spp.), bamboo and rosemary.

The herbaceous layer At up to 1m (3'), these are plants that reach a good height but mostly die down in winter, unlike the shrubs with woody stems. This is a varied layer that could include plants such as artichokes, asparagus, cardoon, fennel, perennial kale and broccoli, comfrey, giant butterbur (*Petasites japonicus*), crimson clover, vetches, herbs such as mint, sweet cicely and lemon balm, and even yacon and oca (see Chapter 4 for more about perennial vegetables).

The ground layer These plants cover and protect the soil. Here you could grow low-growing plants such as Nepalese raspberry (*Rubus nepalensis*), ferns, wild onions and leeks, wild strawberries, buckler-leaf sorrel and white clover.

The rhizosphere This is the root zone, rich in soil life such as bacteria and fungi. Mycorrhizal fungi form mutualistic relationships with the roots of many of the plants. Their fruiting bodies, the mushrooms, appear above ground in autumn.

Climbers These plants root on the ground and clamber up into the canopy. Examples include blackberries, hops and yams.

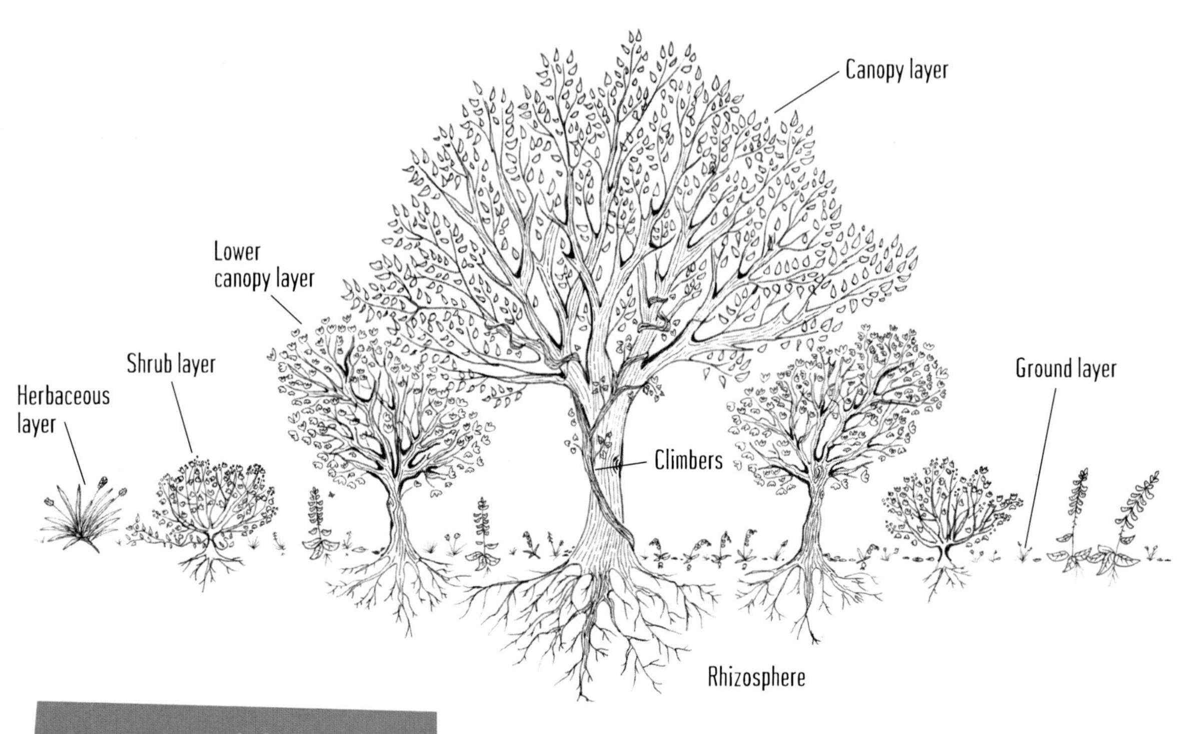

The seven layers of a forest garden.

The rhubarb-like large leaves of the Japanese or giant butterbur (*Petasites japonicus*) thrive in the dappled shade.

Bamboo

Establish a clump of bamboo on your plot and you won't need to buy bamboo canes ever again! Bamboos are edible too. The young shoots are harvested before they are 2 weeks old, so they are still tender. They are then peeled and cooked for 20 minutes before being used in stir-fry dishes, etc.

There is a variety of bamboo species to choose from – but beware that some are more invasive than others! Their spreading rhizomes can pop up metres away from the parent plant, so for these species you need to put in a physical barrier such as a root-barrier membrane before planting. Bamboos of the genus *Fargesia* are clump-forming, so lack the invasive rhizomes. They are shade-loving and very hardy. The umbrella bamboo (*Fargesia murielae*) is very attractive, with yellow canes up to 4m (13') high. Bamboos of the genus *Phyllostachys* are tall, vigorous and spreading, with thick canes up to 6m (20') high. They can cope with full sun or light shade, and are great for screening.

Grow your own spices

If you do a lot of pickling and cooking with spices you will know how expensive it is to buy just a small pack of spices. The good news is that a surprising number of hardy shrubs and trees can be grown to provide spices. For example, a number of plants of the genus *Zanthoxylum*, from Nepal, China and Japan, are great substitutes for black pepper. The best known of these is the Szechuan pepper (*Zanthoxylum schinifolium*): a shrub that likes a sunny position, where it will grows up to 4m (13') high. (It is pictured on page 119.) If you like ginger, try *Zingiber mioga* (Japanese ginger), a shade-loving relative of the commercial ginger. Its leaves, shoots and flower buds have a gingery taste and are widely used in Japanese cuisine.

Allspice is a useful spice, especially if you like Caribbean or Middle Eastern cuisine. The main source is *Pimenta dioica*, which is not hardy, but there are some hardy alternatives, such as the wild allspice (*Lindera benzoin*), which grows to 6m (20'), and Carolina allspice (*Calycanthus floridus*), which reaches 3m (10'). Both have aromatic leaves and bark. You can grow your own liquorice too. Liquorice (*Glycyrrhiza glabra*) is a small shrub with edible roots, and it's also a nitrogen fixer.

Planning a forest garden

Planning is critical before planting a forest garden, as space is limited, even with a half-acre or one-acre plot. Of particular importance is the positioning of the trees. Shrubs can be dug up and moved, but not a tree. Make sure you consider the shading and sheltering effect of a tree when it is full grown, and the extent of its roots. Given the space limitations you will need to focus on the smaller tree species and shrubs with edible fruit or nuts (see Chapter 5), as well as edible perennials (see Chapter 4).

It is also important not to dig or rotovate an area destined to become a forest garden, as this will expose the soil and just invite weed seeds to germinate. It is far better to lay down a mulch to kill the ground layer of weeds, and then plant through it. Don't forget to plan the paths and other key structures early on too.

Some people develop a forest garden from an existing orchard. It's not difficult to add a shrub

A mulch of cardboard to kill the ground layer of weeds before planting.

This orchard on a terrace in Hong Kong has been underplanted to create a forest garden.

layer of currants, gooseberries, hazel and hawthorn, and some climbers such as roses, raspberries and blackberries. Some of the existing grassy ground cover can be mulched and planted with perennials such as artichokes, mints, hostas, ostrich fern (*Matteuccia struthiopteris*), walking onion, chives and violets. Even dandelions, with their deep roots, can be valuable.

Alternatively, a large poultry pen could be a good starting point. For example, plant a couple of fruit trees and some perennials that are robust enough to cope with the birds, such as comfrey, to create ground cover. The benefits are quickly realized, as the trees provide the poultry with shade and windfalls, while you get a second crop from the same area.

PART THREE

KEEPING LIVESTOCK

chapter SEVEN

Poultry for eggs & meat

The chicken is probably the most popular animal to keep on a smallholding, but it's not the only bird to consider. Ducks, geese and turkeys can be raised for either eggs or meat, and are ideally suited to small plots. Or there's the tiny quail, which is a prolific egg-layer.

It's not hard to find room to keep chickens: even a limited space should be plenty for both egg-layers and birds raised for meat. Ducks and even turkeys will be happy in a modestly sized pen, while geese are grazing animals, so need plenty of grass. With space at a premium, try to include poultry in your crop rotations or run them under fruit trees, to maximize the use of the land and get 'two crops from one space'.

One of the most popular hybrids is the hardy Black Rock.

Poultry for eggs

If you decide to keep chickens, there are many choices available. This section describes the options, and explains the needs of a laying flock.

Choosing your chickens

Laying hens are usually bought as point-of-lay pullets, that is, birds that are about 18-20 weeks old and about to start laying. When choosing your birds, you have a number of options: hybrid, traditional or 'ex-bat'.

Hybrid chickens

Commercial hybrids can produce in excess of 320 eggs a year, and many, for example the Black Rock or Warren, are bred for a free-range system. Some commercial poultry breeders have bred their own range of hybrid egg-layers, and often the choice of the bird comes down to the colour of the egg that it produces. Although they lay well for a few years, hybrid birds are not as long-lived as a traditional breed.

You can expect a young hybrid bird to lay well for its first year, even through winter, and then production declines through the rest of its life.

Some commercial breeders have bred their own range of hybrid egg-layers, and often the choice of bird comes down to the colour of the egg that it produces.

Our Silverlink hybrids have mostly white feathers with some brown, and lay brown eggs. They are incredibly friendly and docile, and are happy to be picked up.

Older birds still lay eggs – just not very many. Hens moult in the autumn of their second year, and then every autumn, during which laying either declines greatly or ceases altogether for around 4 to 6 weeks. They will lay fewer eggs during winter too, as there is not enough light to stimulate egg laying. You can overcome this by providing artificial lights that come on a few hours before dawn to extend the daylight hours.

Traditional breeds

There is a huge choice among the traditional breeds, such as the Light Sussex, Welsummer, Maran or Buff Orpington. Over the years, poultry breeders have selected birds for their appearance and not for their egg-laying abilities, so these birds can look great, but they may only lay 60 or 70 eggs a year, during spring and early summer. You should therefore check the egg-laying records of the breeder. The darkest eggs come from the Welsummer, Maran and Penedesenca (in the USA), some lines of which lay amazing chocolate-brown eggs that are much desired and very popular with consumers. Also in demand are the blue eggs laid by the Cream Legbar, Araucana and Ameraucana (in the USA).

A trio of Welsummers. The hens lay very dark brown eggs.

Some recently acquired ex-bats in their new pen.

Ex-bats

An alternative to buying point-of-lay birds, either hybrid or traditional, is the 'ex-bat'. This is a commercial bird at the end of its (commercial) laying life. Commercial birds are not kept beyond the moult (around 78 weeks), so they are usually slaughtered at that point, although they could go on laying for several years. Nowadays, the battery-cage system is no longer in existence in the EU, so ex-bats come from colony cages (with more space than a battery cage), or even from a free-range flock. The birds will look a bit feather-bare as they are moulting, but this appearance does not mean that they have been mistreated.

Moulting hens look a bit feather-bare but the feathers soon grow back.

These birds require extra care during their first few weeks as they adjust to a new system. If they have been housed inside with lights, they will need to adapt to a free-range life, as they won't be used to returning to their house at dusk, using the nest box or even perching. They can be nervous at first and not wander far from the house, but they will soon learn to scratch or explore. They will need a good-quality food to help them regrow their feathers, plus a vitamin-and-mineral supplement in the food or water. But with a bit of care and attention these birds can provide several years' worth of eggs, so they are an option for people who like the idea of giving a home to birds that would otherwise be slaughtered.

Introducing new hens

Problems can crop up when you introduce new hens to an existing flock, especially if they are much younger, or there are just one or two. When introducing hens, always place them in the hen house so they can come out in their own time and know their way back. Put them in the pen in the open and you'll be chasing them around at dusk!

The best time to introduce the newcomers is at night, as hens rarely squabble at night. Pop them in the house, either on the floor or on an unoccupied perch, and leave them. In the morning, the flock will rush outside, leaving the newcomers, who will take their time. Outside, it's best to make sure that there are several sources of food and water, as the bullies in the flock are quite likely to chase the newcomers away from the food. You can also hang some sunflower heads, a cauliflower or a cabbage in the pen to distract them.

Hens love fruit and seeds, so grow sunflowers around the plot to provide a supply of sunflower seeds.

A good feed

Laying hens need a good-quality feed, such as a mash or pellet that has been formulated for their needs. An adult bird will require 50-120g (approx. 2-4oz) of food a day, depending on its size and the time of year. Wheat or maize can be added as energy sources in winter, especially in the afternoon to keep them warm at night, but remember that these foods are a bit like chocolate to a chicken! You can also give treats in the form of greens, seedheads, dandelions, chickweed, groundsel, etc.

To prevent wild birds and rats snacking on the food you can use an automatic feeder, which either releases a small quantity of food at a time or is operated by a treadle on which the birds stand to reveal the food.

In addition to feed, all poultry need grit for their digestive system, and oyster shell as a source of calcium.

REMEMBER that it is illegal in the EU to feed chickens any food that has been in a kitchen or catering establishment, whether domestic or commercial (see Appendix).

Our small flock of hybrid hens has a plastic house with six nest boxes, and a shelter in their pen made from pallets.

Housing

Hens roost at night, so they need a house or shed. If your pen is not completely predator-proof, they will need to be locked in every evening for their own protection and let out again in the morning. Most hen houses are made of wood, but there are also houses made from plastic and recycled plastic sheets. These have a number of advantages over wood, especially when it comes to the control of the dreaded red mite (see page 157).

When making or buying a house, ensure you have enough space for the number of birds you intend to keep. You need to allow a minimum floor space of 30cm x 30cm (12" x 12") per bird, that's about nine or ten birds per square metre (11 square feet), and if you prefer higher welfare standards then aim for the organic standard of six birds per square metre. However, do not have too few birds in a house, as they huddle together and their body warmth keeps them warm in cold weather. If there are too few birds in the house, they can get cold. Ventilation is important too, so look for vents that will provide a good flow of air. Many poultry keepers are tempted to close up the ventilation holes in cold weather to keep the birds warm, but ventilation is essential for their good health.

Hen houses are heavy, especially the larger ones, so think carefully about where you will position it. It needs to be oriented so that the entrance is facing away from the prevailing wind. If the site is windy, provide a windbreak in the form of a screen, mesh or even a straw bale to give the birds shelter. If you are rotating your birds on to fresh ground regularly, you will have to move the house from pen to pen, so you will need skids, wheels or carrying handles. A raised house with space underneath provides shelter from the rain and sun.

Hens also need a perch to roost at night. A perch is a rod of wood or plastic with a diameter of 4-5cm (1½-2") so the bird can wrap its feet around it to grip on. Space-wise, allow 15-25cm (6-10") of perch space per bird. Another requirement is nest boxes: at least one box per three birds, lined with nesting material, for example wood shavings, straw or hemp bedding. The pophole (entrance) of the house needs to be large enough for the birds to get out easily and needs a fastening that prevents the fox from getting in. A layer of bedding should be spread over the floor to take the dirt off feet and catch droppings.

The poultry pen

Your hens need to be contained: if left to range free, they are not only at risk from the fox but are also likely to demolish any vegetable beds. There are a number of options to consider. On a small plot, it is feasible to erect fox-proof fencing around the perimeter of the whole site (see Chapter 1, page 30). Although expensive, this could be a good investment. My own plot has perimeter fencing with tall stakes supporting three lines of wire to deter the deer, and two electric lines near the ground to stop rabbits and foxes getting in.

My poultry pens are fenced using panels, each about 3m (10') in length and 1.8m (6') high, constructed from a wooden frame and chicken mesh (pictured below). These panels are easy to make and erect, and are supported in place with

These poultry pens are easily erected and dismantled, so they can be moved regularly.

Placing a shelter in the poultry pen provides the birds with shade and helps to draw them away from the hen house.

The wild ancestor of the hen, the jungle fowl, hunts on the forest floor for insects, leaves, shoots and seeds.

a wooden stake and plastic ties. To secure this against the fox, an electric line could be run along the top and bottom of the fencing. Another option is electric poultry netting, which consists of a plastic net supported by plastic stakes, and a battery unit. This is an easily moved system (see Chapter 1, page 24).

Hens don't need a huge amount of space, but if you put too many in a small area they will quickly remove the vegetation, leaving your soil exposed, so every time it rains you get a muddy mess and dirty eggs! Then when you move the pen, the bare soil is quickly colonized by weeds rather than grass, and you end up with a weedy patch. Ideally, you should aim to provide at least 10m^2 (108 sq ft) per bird, which should be sufficient to ensure good cover year-round and give the hens something interesting to investigate, where they can scratch and find things to eat.

In the wild, the ancestor of the domestic hen, the jungle fowl, hunts for food on the forest floor, feeding on spiders, beetles and other insects as well as leaves, shoots and seeds. One way to encourage a rich diversity of invertebrates is to allow the grass to grow long and to incorporate flowers that attract insects, such as yarrow (*Achillea millefolium*). Such a meadow-like habitat could supply up to 20 per cent of the birds' nutritional needs. Alternatively, you could provide more of a forest garden for the poultry, similar to their original jungle environment, by planting trees and shrubs (which would provide leaf litter for them to scratch) and including log piles to attract beetles and spiders. The trees will also offer shelter from the weather and protection from aerial predators. A willow coppice grown for fuel or a nuttery would also provide an excellent habitat for hens. (See Chapter 6 for more on tree crops and forest gardens.)

If you have less space, then you will have to move the pen regularly to keep the ground covered. Don't forget to make use of your perennial vegetable beds, orchard and fruit cages during the autumn and winter months, as the birds can range under the shrubs and trees, gaining shade and eating any fallen fruits, but not damage the plants. If you only have a few chickens in a house and run, you can rotate that around your vegetable beds, so that the chickens clear the beds of pests before being moved to fresh ground.

Hens, like all birds, suffer from intestinal parasites, such as roundworms and tapeworms (see page 158), so the other important reason for rotating pens is because it will reduce the worm burden. The eggs of roundworms can survive in

Going broody

A broody hen is one that sits on her nest, even when it's empty, and pecks at you when you try to reach the eggs underneath her or move her off the nest. It's not such a problem in hybrid hens, as they have been bred not to be broody, but it is common in the rare breeds and particularly in bantams such as Silkies. It becomes an issue, as a broody hen will sit all day and may not feed or drink, even in hot weather. Broodiness can last for 2 or even 3 weeks, and if you allow her to sit she may die of starvation. Remove her from the nest and put her outside, closing the pophole if necessary to stop her slipping back. Check that she has taken some food and water before opening the pophole. Never encourage her by feeding her inside the house or leaving water beside the nest box. If this doesn't work, you may have to remove her and keep her in a cage away from the flock until the broodiness passes. Of course broodiness can be useful at certain times of year, for example if you wanted to hatch a few eggs, and Silkies are excellent mothers.

A broody hen refusing to leave the nest box.

Duck eggs have a richer taste than hens' eggs, because of the larger yolk, and also higher levels of vitamins and minerals. They are favoured by bakers for sponge cakes.

the ground for at least 3 months, especially in shady places and in the damp areas around water troughs, so the worms' life cycle can be broken by resting the ground for several months, or longer if possible. Once the ground has been used, move the birds to a fresh pen, rake the surface of the old ground and lime it to adjust the pH of the soil, as chicken droppings acidify the soil. If the ground cover is patchy, consider re-seeding with a poultry seed mix with grasses that can withstand the scratching of hens.

A chicken scavenging system

In many countries, chickens roam around settlements and exist on waste domestic food. While this is not possible in the UK and many other parts of the world, due to rules regarding catering waste (see Appendix), there is no reason why chickens should not be allowed to pick over the weeds and other green waste from the plot. Their pen could be used as a compost heap, and you could drop off barrowloads of green waste for the chickens to scratch over. Their faecal deposits aid the breakdown of the waste, while the abundance of invertebrates and seeds provides part of their diet. Once enough waste has built up, the birds can be moved to a new pen and the process repeated.

Speciality eggs

Hens are not the only birds that can be raised for their eggs. The eggs from ducks, geese, turkeys, quail, pheasants and rheas are all options. Quail eggs (see page 144) have long been sold on supermarket shelves, as have duck eggs, and while the others are more of a novelty, they are just as tasty. In North America, turkey eggs were widely eaten until the turn of the twentieth century, when they fell out of fashion.

Duck eggs are becoming increasingly popular, especially in North America and among Asian communities. Duck eggs have a richer taste than hens' eggs because the larger yolk means there is more fat present. They also have higher levels of vitamins and minerals. Many people who are allergic to hens' eggs are not allergic to duck eggs. Duck eggs are favoured by bakers, who use them in sponge cakes for a richer, more moist texture, and are appearing more frequently in chef's dishes. The two main laying breeds are the Khaki Campbell and Indian Runner, which lay between 150 and 200 eggs a year. These breeds are behind the modern hybrid duck, which can lay in excess of 250 eggs a year.

Like laying hens, egg-laying ducks must have a good-quality diet. A Khaki Campbell, for example, needs about 140g (5oz) of food a day. Like hens, ducks are affected by decreasing light levels, and egg production falls off in winter unless artificial lighting is used. They are productive for a couple of years and then egg laying decreases. Ducks tend to have longer lifespans than laying hens.

An alternative is geese kept for eggs. A good laying strain will produce an egg every other day for 3 months from late winter to spring. The large eggs are perfect for baking, and if you sell them you will soon recoup the cost of keeping geese. The eggs weigh around 180g (6oz). For more about geese and ducks, see page 150.

Turkeys are seasonal too, laying up to 100 eggs each year, with a laying season of as much as 20 weeks. The speckly brown eggs weigh around 100g (3½oz), and although they taste like hens' eggs they contain more fat and cholesterol. See page 154 for more on turkeys.

Large birds such as rheas will only lay a couple of dozen eggs a year, but they are extra-large. The larger eggs of the goose and rhea are also in demand from people who like to paint egg shells, so there can be a good market for their eggs.

Quail

Quail are small birds that are well suited to the small plot. They are great egg-layers and very easy to keep. Quail eggs are in demand and often more expensive than the hen egg, so a small flock of hen quail would soon pay for itself. The most common variety is the Japanese (or Coturnix) quail, as it is fast-growing and can be sexed at around 3 weeks, while other varieties include the Northern Bobwhite and California quail. Quail are also raised for meat, which is popular, and there are some docile hybrid lines bred

The quail is surprisingly productive, laying up to 300 eggs a year.

especially as meat birds. Unless you are raising birds for meat, you should keep only adult hen birds or hen birds with one male, as the males are likely to fight.

Quail housing is simple. They do well in a modified rabbit hutch attached to a large run. While rabbits like shady resting places, quail need more light to encourage laying, so remove the back of the hutch and replace it with mesh. A hutch or cage of about 100cm x 60cm (39" x 24") would make a house of sufficient size for about six quail. The birds will need a nest box at the back of the house or sited within the pen, and the floor of the house and nest box should be covered with nesting material such as straw or hemp bedding, with an extra-thick layer in winter. Quail are not as hardy as chickens, so in winter you will need to provide some form of heating in their house or move them into a garage or shed, or keep them in a polytunnel. The birds like to perch, so you should provide perches both in their house and in the run. Place the feeder and drinker inside the house. If the run is secured against the fox there is no need to lock them up at night, as they will move themselves into the house if necessary. Their run needs to be placed in a sheltered position out of the wind and shaded from sun and rain, and it also needs to have an area for dust bathing, as this is an important natural behaviour for them.

The quail is a good flier, so the run needs a roof to contain them, even if they seem quite tame. Care needs to be taken when opening up their house or run, as an escapee may never be seen again! If startled, quail tend to fly straight up to escape danger, and this means they can fly into the roof of their pen, which if made of mesh could injure them. This risk can be reduced by adding an inner roof of soft netting such as that used to keep butterflies off crops.

Eggs can be hatched in an incubator or under a broody hen, and the chicks moved to a brooder for about 5-6 weeks, then they can go outside. (In many respects the process is the same as for raising chickens – see page 148 for full details.) There are some specialist game feeds around, the same as those fed to partridge and pheasant, but if you only have a small number of birds, chick crumbs are fine. Once the birds are moved outside, those intended for egg laying can be fed a layer's pellet, while birds that are being raised for meat can be fed a grower's pellet. As well as their feed, quail will forage for fruit, grain and small invertebrates which they can find in their run, so make sure you allow the grass to grow and have a few low-growing shrubs for cover, to provide insect habitat.

The hens come into lay at just 5-6 weeks, and a good laying strain will produce as many as 300 eggs a year; during the long days of summer the hens may even lay two eggs a day. The females live for about 3 to 4 years and the number of eggs laid falls by half in their second year and then continues to decline.

Raising chickens for meat

Most smallholders think of keeping chickens for their eggs, but many forget that they can be kept for their meat too. Chicken has been eaten for thousands of years and was a common meat during the Middle Ages. At that time the birds were dual-purpose: the hens for their eggs and the cockerels for their meat. At the end of their laying life the hens were slaughtered and used as boiling fowl. Today, the term 'meat chicken' describes a chicken bred specifically for meat production. It grows quickly, with a good conversion of feed into muscle, produces plump breast meat and can be slaughtered within months.

This batch of meat birds came from a Rhode Island Red x Light Sussex cross.

Having a freezer full of chicken that you have raised yourself is very satisfying. Not only do you have meat from birds that have lived free-range, but you know that they have had a good-quality life and a quick end. The good thing about raising meat birds is that you don't have to keep the birds all year round. Instead, you need only 12-14 weeks and a small pen. In fact, a pen of just 14m x 17m (46' x 56') would be plenty large enough for a batch of up to 30 meat birds. This could be an open grassy pen or an area of orchard or even a willow coppice. Just remember that each batch of chicks needs to go out on to clean ground that has not been used for poultry for a year, to reduce the risk from parasites (see page 157). With careful planning, the meat birds could be part of the land-use rotation with vegetables, pigs and poultry.

Choosing your meat bird

As with egg-layers, there is a wide selection of both commercial hybrids and traditional breeds for meat production. The main commercial hybrids are ready for slaughter in 7-15 weeks.

Cornish cross This is a popular choice in North America, as the birds are very hardy and will forage far from the house. They reach 2.4kg (5lb 5oz) in 8 weeks.

Hubbard There are various strains, including the Hubbard Coloryield and JA57, with regional names such as Poulet Bronze, Mastergris and Farm Ranger. The Hubbard is slower-growing than the Ross Cobb (see below).

Ross Cobb This is the commercial fast-growing white-feathered meat chicken, reaching up to

The slower-growing Mastergris is a strain of Hubbard.

a massive 3.5kg (7lb 11oz) in 12 weeks. Commercially, it is slaughtered at 6 to 7 weeks. It is better suited to barns than free-ranging, and is prone to leg problems due to the fast rate of growth and heavy weight.

Sasso This is a French gourmet chicken. There are various strains, which vary in their feather colour, skin and leg colour, etc. They are all bred for slow growth, producing healthy birds that are suited to free-ranging, with excellent-tasting meat.

There is also a number of traditional heavy breeds that are well suited to meat production, for example the Rhode Island Red or Dorking. These birds are slower-growing and may not reach weight until they are 20-26 weeks old. Most are dual-purpose breeds, as the females can be used for egg laying.

Cornish This breed was the foundation of the broiler (meat bird) industry. The development and arrangement of its muscles gives it a good carcass shape. The skin is yellow and the bird's weight can reach 4kg (8lb 13oz).

Dorking This is known for its fine quality of meat. It is slightly smaller than the Rhode Island Red, finishing at 3kg (6lb 10oz). It has a rectangular body and short legs.

Ixworth An old traditional breed, this has a good compact shape and an excellent flavour. It takes 20 weeks to reach size, and is dual-purpose.

Langshan These are available in two varieties, one white and the other black. The standard weight varies from 2.5kg to 4kg (5lb 8oz to 8lb 13oz). It is generally used for meat rather than eggs. It's a very active and energetic breed.

The Rhode Island Red was created as a dual-purpose breed, with the cockerels raised for meat.

Rhode Island Red Weighing around 3-4.5kg (6lb 10oz to 10lb), these too are dual-purpose. The best strains lay over 250 eggs a year.

Day-olds or hatching eggs?

Most meat chicken producers supply day-old chicks. If you have a suitable incubator, you could order hatching eggs, which need to be incubated for 21 days. The day-olds need to be kept under heat in a brooder for 3 to 4 weeks until their feathers develop and they are ready to go outside. A brooder is a simple arrangement that you can make yourself (see box below) with a heat source such as a suspended infrared lamp or an 'electric hen', which is a heat mat on legs. The chicks squeeze under the heat mat, which mimics the way the mother hen would keep her

Making your own brooder

This is a circular brooder. Chicks can be quite skittish, and when alarmed they can pile up in a corner, with a risk of getting squashed, so a circular design avoids this. You will need two lengths of hardboard or stiff card, about 30cm wide by180cm long (12" x 70"), which are bent round and held together with two wooden pegs, creating a circle about 90cm (36") in diameter. Place newspapers on the floor and top them with a thick layer of dry wood shavings or hemp bedding.

Add the heat source, such as an 'electric hen' or an infrared bulb. The height of this needs to be adjusted to maintain the right temperature for the chicks. If the chicks are too hot, they will move to the sides of the brooder and may be panting, in which case you should extend the circle to give them more space to spread out. If they are cold, they will sit directly underneath the heat and may be clambering on top of each other. If this happens, lower the heat source a little and watch them closely. With the ideal level of heat the chicks should be settled and sleeping without being directly under the heat source. It is important to enlarge the brooder as the chicks get bigger, as overcrowding is one of the main causes of chick mortality.

An 'electric hen' is a heat mat on legs, which mimics the way the mother hen would keep her chicks warm. As the chicks get larger, the heat mat is raised.

chicks warm. As the chicks get larger, the heat mat is raised. I prefer the electric hen to a heat lamp, as there is less chance of the chicks overheating and it avoids the risk of the lamp failing. The brooder may be housed in a shed or a utility room, or even a garage. The brooder temperature should start high and be gradually reduced over the following 4 weeks. For example: 1 day old, 35°C (95°F); 1 week, 33°C (91°F); 2 weeks, 30°C (86°F); 3 weeks, 28°C (82°F); 4 weeks, 25°C (77°F).

Chick feed and care

Day-old chicks are fed 'starter crumbs', which have a high protein content.. Young chicks are always offered unlimited quantities of food to make sure their growth is not checked. The crumbs can be placed on an egg box, which can be thrown away when it gets dirty, and you can switch to a proper feeder after a few days. By the time the chicks are 3 to 4 weeks old they can be moved over to grower's pellets, which are also high in protein. However, if the young birds grow too quickly there is a risk of them developing leg problems. If you think this is happening, switch them over to a layer's pellet, which will slow down their growth rate. On average, a single bird will eat 5-6 kg (11lb to 13lb 3oz) of grower's pellets by the time they reach slaughter weight. Some chick crumbs contain medication for protection against coccidiosis, a parasitic infection often associated with damp litter or over-used ground (see page 158). This is fine for chickens, but do not use medicated chick crumbs for ducks or geese.

The chicks will need a supply of fresh water, which can be either dispensed using a small chick drinker or provided in a shallow dish, partly sunk into the shavings and with a layer of pebbles to stop the chicks from paddling.

It is essential to provide the chicks with a clean and friable bedding, such as sawdust, hemp or chopped straw. Wet bedding soon becomes caked, and it will harbour pathogens and generate more ammonia.

Moving to the range

The brooding stage is complete by 3 to 4 weeks, by which time the birds will have feathered up and be able to go outside. By now they will be on a grower feed. Meat birds, like layers, need to be locked up at night, so, depending on the numbers you are raising you will need a shed or house. The house should be low on the ground with a wide, gently sloping ramp and a large pophole for when they reach full size, and they will need plenty of space on the floor to rest. As a general rule you should provide 30cm x 30cm (12" x 12") per meat bird, or about 10 per square metre (11 square feet). Meat birds do not always perch, but ideally perches should be supplied.

Depending on the location, the pasture and house needs to be surrounded by fox-proof fencing or electric poultry fencing. Small mobile houses are good, as they can be moved on to fresh pasture for each new batch of chicks, so with this approach you need to plan on having enough pasture to rotate the different batches.

Slaughter

In many countries, on-farm poultry slaughter is allowed for producers of small numbers of meat birds. This means that if you are raising birds for meat you can slaughter them on your holding so long as you have the appropriate licences in place and you dispose of any waste correctly (see Appendix for more on slaughter regulations). The most common method of home-killing birds – dislocating the neck – is no longer approved by the Humane Slaughter Association, as the bird does not die instantly. This method can be used to kill the occasional bird, but not a whole batch.

If you are going to slaughter poultry on your holding, you will need a plucker and stunner (in the EU a licence to slaughter is needed to use this equipment). The easiest way to slaughter birds is to use the stunner to render the bird unconscious and slit its throat to bleed it. Then it can be plucked. If you are thinking of killing ducks and geese, you will need a wax bath as well, since the down feathers need to be removed. This is done by dipping the plucked bird in hot wax and then cold water and letting the wax set. When it's cold, the wax is removed along with the downy feathers. For killing and preparing turkeys, you will need a sinew remover to remove the sinews in their legs, which are tricky to pull out. The flavour of most birds is improved if they are hung in a chiller for a few days to a week before they are eviscerated (gutted).

If you don't want to slaughter your birds yourself, then there are specialist poultry abattoirs that may be able to take them, although they may require you to have a minimum number.

Ducks and geese

Ducks and geese are gaining in popularity. Ducks are kept either for egg laying (see page 143) or for meat. 'Table birds' (for meat) include breeds such as Aylesbury, Pekin, Rouen and Saxony. There are also some hybrid strains of Aylesburys and Pekins that can reach a weight of 4.4kg (9lb 11oz) in just 10 weeks. The Muscovy is another option. It is a different species to the domestic duck, and the commercial ones are known as Barbary ducks. Ducks are not seasonal like geese and turkeys, so ducklings can be bought as day-olds all year round.

Geese too may be egg-layers (see page 144), while those for the table are mostly white strains based on the Embden, such as the Wessex and Legarth. Day-old goslings are bought in spring and put under heat for about 3 weeks, by which time the downy feathers will have been replaced by waterproof feathers and the young birds can be allowed out to pasture. They take 22 to 28 weeks to reach slaughter weight.

Ducks, and geese especially, will need a larger house or shed than chickens, and they do not perch. Unlike chickens, they do not usually go in on their own at night, so they have to be rounded up and locked in. A small amount of food at night is a good way to encourage them to go inside. As with other poultry, they require security in the form of fox-proof fencing or electric fencing. Most cannot fly, although they may flap over low fencing.

Most people call a heavy white duck an Aylesbury duck, but the pure Aylesbury is rare. Most, like these ducks, are commercial hybrids.

These geese range free during the day and come in at night.

Ducks and geese will quickly make a pond muddy, so a better option is a 'paddling pool' that can be refilled.

Being waterfowl, ducks and geese need clean water to dip their heads. A pond is an option, but they quickly make water muddy, so it's best to have a small 'paddling pool' that can be tipped out and refilled. Shallow auto-filling baths are ideal, as they can be tipped out and left to fill while you do something else. Otherwise various containers can be used.

Space-wise, ducks can be kept in pens of a similar size to that required by chickens, or they could range free in an orchard. Geese need a larger free-ranging space, with plenty of grass. They can be run under an orchard, but make sure that any young trees and low-hanging branches are protected from their beaks! They are good at ring-barking unprotected young trees. Alternatively, if you have a paddock area for a few sheep, you could run your ducks and geese with the sheep.

These commercial white geese are bred from Embdens.

Feeding ducks and geese

Both ducks and geese being raised for the table need the right food in order to reach a good weight. Ideally, ducklings and goslings should be given a specialist waterfowl crumb, which has a different balance of nutrients from the chicken crumb, but if this is not available an unmedicated chick crumb would suffice (see page 149). Goslings in particular need greens along with their crumbs, so a grass turf or dandelion leaves can be supplied. After 4 weeks, ducks and geese can be moved on to a specialist duck and goose grower's pellet, which has a lower protein content. If they are growing too fast, cut back on the amount of pellets you give them.

The ducklings can be left on grower's pellets until they are slaughtered. A hybrid Pekin duck bred for the table has the fastest growth rates of all the ducks. Under ideal conditions, it can reach a weight of 3.7kg (8lb 2oz) in just 7 weeks.

Ducks retained for breeding, egg laying or simply as pets require a daily ration of layer's pellets

Rare-breed geese such as these American Buffs can be kept for breeding, as they are in demand from poultry enthusiasts.

The large exhibition Toulouse is another rare breed that sells well.

(for laying hens), which they will supplement with slugs, snails, worms and other animal life that they find in the grass and mud.

Goslings stay on grower's pellets for about 12 weeks and can then be moved to a wheat diet for summer. By autumn they are eating about 200g (7oz) of food per day. The final weight of the bird is 5-9kg (11lb to 19lb 13oz), which, when plucked and dressed, produces a dressed weight of 3-7kg (6lb 10oz to 15lb 7oz).

Geese are grazing animals, so any adult geese on your plot need a constant supply of grass, and when this is available during the summer months they do not require any other food. At other times they can be fed a layer's pellet, wheat or corn. If you intend to breed your geese, move them on to a specialist breeding-goose ration in late winter if possible; otherwise feed them a layer's pellet.

Turkeys

Turkeys are excellent birds: impressive, chatty, friendly and inquisitive, and easy to keep, so they are a good addition to the holding for either meat or eggs (see page 143). As with other poultry, there are a number of turkey breeds available. The commercial white turkey is a fast-growing bird that is better suited to barn-raising than free-ranging. More suited to the smallholder are traditional breeds such as the Bronze, Norfolk Black, Bourbon Red, etc. These breeds are hardier, the meat is much more tasty, and they don't get quite so heavy, which means the birds continue to be quite active in their last weeks. You can buy turkeys as day-old chicks or as 4-6-week-old poults in late spring and early summer. Some suppliers of Bronze turkeys offer a selection of strains, each strain reaching a different final weight, which enables a small producer to produce a range of table birds.

Turkeys range during the day and require a shed or barn for roosting at night. An adult turkey is a large bird, so you need to allow 1m^2 (11 sq ft) per bird: for example, a 2.4m x 1.8m shed (8' x 6') gives 4.1m^2 (44 sq ft), enough for four adult birds or eight to ten birds raised for Christmas. The shed or house needs plenty of ventilation. If there is a window, you can remove it and replace it with mesh to keep pests and birds out, or cut out a window-sized hole in the wall and cover with mesh. Just make sure the hole is not facing the direction of the prevailing wind.

Although young poults are active and happily fly up to a perch, there is a risk of them denting their breastbone, so they should be given straw bales instead and only allowed to perch from 16 weeks. Adult turkeys need a sturdy perch positioned at least 70cm (28") from the ground. They

The Norfolk Black turkey looks impressive but doesn't reach the weight of the Bronze turkey. It has a gamey flavour.

can fly up to a perch in a small space with relative ease, but have more difficulty getting down and may crash-land! This means there is a risk of damage to the legs and feet, and they can get a condition called bumblefoot, caused by a *Staphylococcus* infection. As the birds approach their kill weight they may be too heavy to perch, so the perches can be removed to give them more space.

The turkey pen needs to be well fenced, as these birds are escape artists. A livestock fence is no barrier to them, so you need netting of at least 1.8m (6'), or place pheasant netting over the pen. Space-wise, a trio of adult birds will be happy on a 10m x 5m (33' x 16') grassy pen. If you are keeping birds permanently, then try to have at least two pens, so you can rotate your birds between them.

Raising turkeys

If you buy day-olds rather than 6-week-old poults, they need to go in a brooder under heat for 4 to 5 weeks. The young chicks do not have great eyesight and they are easily spooked, so you need to make sure that they can find their food and water. Some people hang lights; others will use shiny objects like marbles in the water to grab the chicks' attention. While they are inside and under heat, the chicks are fed a turkey starter crumb, which has a higher protein

A batch of Norfolk black turkeys being raised for meat.

content than a chicken crumb. Then they are moved over to a turkey grower's pellet until they are 10-12 weeks old. At this point they are fed a turkey finisher pellet until the last few weeks, when they are switched to oats for fattening. While foraging, turkeys will supplement their diet with grass and small insects.

If you are raising the turkeys as meat to sell at Christmas you need to think about how you are going to tackle the slaughter, plucking and dressing of the birds. Selling them means that you are subject to more rules than if you're just raising them for your own consumption. Being much larger birds than chickens, they are more difficult to kill, and you have to stun and bleed them if you intend to sell the meat (see Appendix). The best option is to have them killed at a poultry abattoir, but you need to book early as it's a very busy time of year. If you can solve the slaughter challenges and have a market for your birds, raising a small batch of turkeys for the Christmas market can be very profitable.

Disease in turkeys

If you are keeping both turkeys and chickens or quail on your holding, you have to be careful where they run. Turkeys are affected by a parasitic disease called blackhead or histomoniasis, for which there is no medication available in the EU. It is caused by a parasitic protozoan carried by intestinal worms found in chickens, turkeys and game birds, so it's a parasite within a parasite! The parasite is found in the eggs of the worm, so it is picked up by turkeys and

Take time to observe your birds each day, so you can spot problems as soon as they appear. Be aware that wild birds can harbour disease, as can rats and mice.

chickens when they scratch around in the earth. Earthworms may also carry the disease. Infected eggs can remain in the ground for up to 5 years. Although chickens and game birds carry the parasite, they are not affected by it, but turkeys are. Symptoms include yellow droppings and darkening of the head. To prevent the disease, make sure turkeys do not run on ground that has housed chickens (or game birds, for example quail) in the previous 5 years. If you have to rotate turkeys with chickens, the best preventative measure is to worm them every 6 weeks.

Parasites, diseases and other problems

Poultry are affected by a number of parasites, both external and internal, plus they are susceptible to some common diseases. The key to keeping healthy birds is knowing how your birds behave when they are fit and well. Take time to observe your birds each day so you can spot problems as soon as they appear (see Appendix). For example, a change in the appearance of their droppings is often the first sign of disease. Be aware that wild birds, especially starlings, crows and rooks, can harbour disease, as can rats and mice. Depending on the type of fencing used around the poultry pen you could net the pens to keep out wild birds. The following sections cover the common parasites and diseases.

Ectoparasites

Mites and lice are common ectoparasites of poultry, and any infestations will result in decreased productivity and reduced resistance to disease. Mites are arachnids and have eight legs, while lice are insects.

Red mites hide in crevices in the housing during the day and emerge at night to suck blood from the birds as they perch. They are small, pale grey to red in colour, and are seen at night crawling over perches, etc. It is vitally important that regular checks are made for mites during the summer months, as wooden houses can harbour thousands of these parasites, making life miserable for the birds. In the worst cases, the birds will be reluctant to roost at night. Mites can survive in empty housing for up to 40 weeks, so it is essential to carry out a good clean prior to use. One of the most effective treatments for red mite is diatomaceous earth. This is a powder formed of jagged microscopic particles that wear away the exoskeleton of the mite, causing it to dehydrate and die. It is dusted all over the house and added to the dust bath. Mites can also be trapped on fly paper or caught in rolled-up cardboard placed in the roof space.

The northern fowl mite and the louse live on birds. The northern fowl mite is tiny; barely visible to the eye. Typically it is found around the bird's vent, causing the skin in this area to look dirty and greasy, but the dirt is actually the pin-prick-sized mites. These mites can infect just one bird in a flock, so it's difficult to know there is a problem until the birds are really ill. The loss of blood weakens the birds and causes them to suffer from anaemia. Lice are found in particular under the birds' wings and around the vent. They lay their eggs near the base of feathers, 'cementing' them in place to form a hard white blob around the base of the feather. Lice feed on scales and feathers, but not blood. They can be treated by rubbing a louse powder or diatomaceous earth into the feathers.

Scaly leg mite causes the scales on the legs to become raised and rough-looking, due to mites burrowing underneath. They can be treated with petroleum jelly (Vaseline): smear it over the legs and rub it into the raised scales. There are also some medicated preparations.

Worms

A number of internal worms can affect poultry, including roundworms, tapeworms, gizzard worms and throat worms (gapes), and they can be treated with wormers. The one that is used most commonly for poultry is flubendazole, which is available in powder or liquid form or in a medicated feed. The powder is added to their feed for 7 days, while the liquid is added to the drinking water. Flubendazole is easy to administer and it is safe to use on geese, turkeys and game birds, but it is not licensed for ducks. The wormer will kill the parasites in the birds, but there will still be thousands of eggs in the ground, so you may need to repeat the treatment or, better still, move the birds to a clean pen. Some chicken keepers recommend that you worm birds regularly, but it is far better to get faecal egg counts done first to see if you have a problem, rather than worm them unnecessarily. Regularly moving your poultry on to fresh ground will reduce the need to worm.

The worm species that parasitize poultry are different from those that affect pigs, sheep and goats, so, for example, you can move pigs on to ground that has had poultry. In fact, poultry can help to reduce the worm load on paddocks used by sheep and goats by eating worm eggs.

Coccidiosis and other diseases

Coccidiosis is the most serious disease affecting poultry, and is caused by a parasitic organism that grows in the gut and passes out in its faeces to be picked up by others. Symptoms include mucky back ends with blood in faeces. Young birds are particularly susceptible at 3 weeks old, when their feathers start to grow. The best way to avoid this disease is to ensure that the litter in the brooder is clean and dry and that the chicks are not overcrowded. Make sure there is clean water and plenty of food so they don't need to scratch around in litter. It is possible to use medicated chick food, but this should only be given to chickens and turkeys, not to geese or ducks.

Birds can suffer from a range of bacterial infections. Symptoms include lack of appetite, weight loss and watery droppings. While birds can become lame, especially the heavier species, a joint that is swollen and hot suggests a bacterial infection, requiring treatment with antibiotics.

Poultry are also affected by a number of respiratory diseases, causing them to have runny eyes and to sneeze, gurgle and have difficulty in

These chickens made straight for the seeds on these dead shoots of cleavers, gathered from a hedgerow.

Natural wormers

Many herbs and spices are known to have anti-worming properties. These include mint (*Mentha* spp.), cleavers (*Galium aparine*), wormseed (*Dysphania ambrosioides*), thyme (*Thymus* spp.), wormwood (*Artemisia absinthium*) and willow (*Salix* spp.). Seeds of cucumber and pumpkin have long been used against tapeworms. A good way to get chickens to eat a variety of plants which may have anti-worming properties is to dump weeds from the smallholding in their pens and allow them to peck over the pile, or collect specific plants from hedgerows. You can buy a natural wormer called Verm-X, which comes either as a liquid to be added to the birds' water for 5 days or as specially formulated pellets.

breathing. Aspergillosis is caused by moulds of the genus *Aspergillus*. The number of mould spores in the poultry house can be reduced by cleaning out the bedding regularly to make sure it does not get damp or mouldy, and topping it up so the birds rest on dry straw.

Another disease to be aware of is avian flu. Spreading around the world, this viral disease exists in a number of strains. There are two groups: the highly pathogenic forms and the less harmful. The disease is often carried by waterfowl that have a degree of resistance to the disease, but migratory waterfowl, such as geese and swans, spread the disease around the world. This is a notifiable disease, so you must contact your vet if you have unexpected deaths of large numbers of birds.

Chapter EIGHT

Pigs for meat

Pigs are wonderfully friendly, entertaining and intelligent. They are among my favourite animals and are another essential for the smallholding. They grow quickly, so you need only keep them from spring to autumn, and just a few pigs should provide enough meat to sell to family and friends to cover your costs.

The Berkshire is one of the smaller prick-eared breeds.

The traditional pig breeds are completely different from the commercial breeds, being adapted to live outside.

Once you have your own pigs, you won't want to buy other meat, as pork is so versatile. It's not just for making chops and joints, but also for sausages, pies, bacon and gammon, pâtés and terrines, and for rendering down to make lard and more. Some breeds are particularly suitable for a small plot, and if you buy weaners in spring, you will have pigs ready to slaughter in autumn.

You need to keep at least two pigs, as they are social animals and live in small groups. On a plot of one acre or less, you are limited to how many you can keep. Pigs will dig up the ground, and this gets incredibly muddy very quickly in wet weather, so with a small space it is strongly recommended that you keep pigs from spring to autumn only, and let the ground recover in winter. Pigs are usually bought as weaners, which, as the name suggests, are young animals that have been separated from mum and are feeding on solid foods. They are between 7 and 10 weeks old.

Before you buy your pigs, make sure you have completed all the required paperwork. In the UK, for example, you must have a holding number before you can move pigs on to your land (see Appendix).

A free-range Gloucestershire Old Spots weaner.

Choosing your pigs

Most of the pork on sale in supermarkets comes from commercial breeds that have been bred to produce lean meat. They are docile in nature so that hundreds can be kept in barns under carefully controlled environmental conditions. Their fast rate of growth means they are ready for slaughter by 16 weeks. The traditional rare breeds are completely different, being adapted to live outside. They grow slowly, achieve a larger size, and their meat is darker, fattier and full of flavour.

Gloucestershire Old Spots

Originating in Gloucestershire, the Old Spots is the oldest spotted pig breed, with pedigree records going back to 1885. It is often called the orchard pig, as these animals were traditionally kept in orchards, where they were fed on windfall apples and whey (the liquid waste from cheese-making). Folklore says that the spots were bruises from falling apples!

You can buy registered pedigree pigs, unregistered pigs and crosses. Registered pigs are pigs that have been entered into the herd book and are born of registered parents. In the UK, the British Pig Association (BPA) maintains the herd books for most of the breeds. Litters from registered pigs are birth-notified with the BPA, which means that a pedigree pork certificate can be obtained when they are sold for meat. For example, to sell meat as Gloucestershire Old Spots pork, the pig must come from a birth-notified litter, as the meat has been given special status as a protected food name, Traditional Speciality Guaranteed (TSG), throughout Europe, just like Parma ham and Welsh lamb. Any pig from a birth-notified litter that is kept for breeding has to be registered, but if you are just raising pigs for meat, registered stock is not necessary. In Australia, the Australian Pig Breeders Association (APBA) undertakes the same role as the BPA, but in many other countries, including the USA, the individual breed clubs maintain the pedigree registers.

Most small pig keepers tend to keep a couple of sows and just hire a boar when needed, and they don't register their litters, in which case the weaners are sold as being of a 'type'. In some

Above: Kune Kune.
Left, top: Kune Kune piglets at 6 weeks.
Left, bottom: Oxford Sandy and Black.
Below: Large Black.

cases, the weaners will be crosses, as the breeder uses a local boar rather than find a boar of the same breed. These animals will be cheaper to buy than birth-notified stock.

Which breed?

For a small plot, the best breeds are the Berkshire, the Middle White and the Oxford Sandy and Black, as they grow quickly and can be finished in 6 to 7 months. They are good at converting food into muscle and they don't have a huge frame, so they get to weight quickly. The larger breeds, such as the Gloucestershire Old Spots and Saddlebacks, are slower-growing. They have a larger frame, which is grown first, so getting them finished takes longer. It's not an issue if you have a decent space for them, but it could be if you want to use a smaller pen and have to run them on during a wet autumn.

Common pig breeds

Berkshire	Small black pig with prick ears, white feet and flash on face. Good for early finishing at under 50kg (100lb), runs to fat if left longer. Good flavour to meat, better for pork than bacon.
Duroc	An American breed with auburn colour, tough skin and prick ears. A very hardy animal, moults in summer leaving bare skin, can cope with hot summers. Has a heavy body and good-quality meat. Often crossed with other breeds.
Gloucestershire Old Spots	The traditional orchard pig, ideal for free-range systems. Tough, hardy and docile. White with a few black spots and lop ears.
Kune Kune	A New Zealand breed. Small, docile and very friendly, popular as pets because of their small size. They don't root as much as other breeds. Their growth rate is slow and they are prone to putting on fat, but they can be slaughtered at 10 months. Great for sausages.
Large Black	Large, all-black pig with lop ears. Docile and hardy, good for free-ranging and easy to keep behind electric fencing. Succulent meat.
Middle White	Small white pig with quirky, dished face with short snout. Quick to mature and easy to manage, usually raised for high-quality fresh meat rather than bacon.
Oxford Sandy and Black	A sandy-coloured pig with black spots and lop ears, hardy with good temperament. A good growth rate, finishing in 22 weeks. Good for pork and bacon.
Saddleback	Large black pig with a white band across the shoulders and forelegs. Hardy. Can be used for fresh meat and bacon.
Tamworth	Large pig with characteristic ginger coat, long snout and prick ears. An inquisitive animal which can be boisterous. Hardy and resistant to sunburn. Has long sides and big hams, good for pork and bacon.
Welsh	White pig with lop ears, long body with deep hams. A hardy pig with fast growth and good feed conversion.

Gilt or boar?

The final choice is gilt (female) or boar (male). Often you have no choice, as the breeder may only have one sex left and the decision is made for you. If you are buying birth-notified stock, it is more likely that the breeder will have boars for sale, keeping the gilts back as breeding stock. Boars will grow more quickly and may reach slaughter weight before the gilts. Traditionally, it is the gilts that are kept a bit longer to be used for bacon and hams, and to avoid any problem of the meat having boar taint. Having kept pigs for many years I can say that I have never eaten pork with taint, despite all my boars being uncastrated and some run on until they were almost a year old. However, it is said that only one in every twelve people can actually taste it. Some research suggests that a taint may be linked to diet and husbandry rather than age.

Generally, it is better to have single-sex groups, as more problems arise with mixed-sex groups, especially when they get to 6 months and towards sexual maturity. If you have a mixed-sex group with uncastrated boars and want to run some on to bacon weight, you would be wise to slaughter the boars early to avoid any risk of the gilts getting pregnant.

How much land?

A pair of weaners doesn't need a huge amount of space: a pen of about 20m x 17m (66' x 56') is fine, although you may need more room if you are keeping them through autumn and winter. You also need to rotate the pens, using different ground each year, to prevent parasites building up in the soil. A rotation over 3 years should be sufficient. We have three pens that we use for pigs, one each year; the rest of the time the pens are either under crops or used for poultry.

This pen had been used for pigs and is now under squash.

Once cleared by the pigs, the land is ideal for use for growing vegetables, as the pigs will have fertilized it with their urine and dung. Ideally, cover the soil after the pigs have been moved so the weeds don't regrow. This land is particularly good for hungry crops such as potatoes, brassicas and squash. The following year you can use the ground for legumes or roots, or re-seed it to keep poultry, before returning it to pigs.

Housing and fencing

Outdoor pigs need a shelter. The standard size of arc is 2.4m x 1.8m (8' x 6'), which will accommodate four to six weaners to pork weight. Traditionally, arcs are made from corrugated iron supported on a heavy wood frame, which means they can be difficult to move without the

Young Berkshires with their housing on one of our pens . . .

. . . and the cleared plot 5 months later.

help of a tractor. If you are only keeping a couple of pigs, you could choose something a bit smaller and more manageable, as the arc only needs to be about 1.8m x 1.2m (6' x 4').

The easiest option is to buy a purpose-made pig arc, and there is plenty of choice, although they are not cheap. When the pigs get towards slaughter weight, they are quite strong and can damage poor-quality houses, so look for strong supports for the roof and exterior-quality plywood. Check that none of the timber is treated with a preservative that could harm the pigs if they chew the wood.

These Berkshires are housed in a wood and corrugated iron arc.

Alternatively, you could adapt a shed, but again bear in mind that pigs are strong and that they rub against their shelter. If you are only keeping the pigs during the warmer months, you could get away with a small straw-bale arc to provide shelter from the sun and rain, or build a temporary shelter from pallets and sheets of corrugated iron.

Some arcs come with a floor and others don't. If there is no floor, you need to provide a thick layer of straw to insulate the pigs from the ground. If there is a floor, make sure it's not too slippery and add a few battens to give the pigs some grip. Pigs are surprisingly clean animals, so they will not urinate or defecate inside the arc, but it will still have to be cleaned out regularly, as it can get muddy.

Think about where you are going to place the arc. The main opening should be positioned away from the prevailing wind, ideally in a sheltered position so it gets some shade in summer. Don't forget that an arc with a corrugated iron roof can get very hot in the sun.

Another critical aspect is fencing, and there are two main options: livestock fencing or electric fencing (see Chapter 1, pages 22-5). For small plots where permanent fencing is required, you can erect a livestock fence and run a line of barbed wire or electric wire along the bottom to stop them rooting under it.

Electric fencing is good for large areas, and it is relatively cheap and easy to move. However, young pigs need to get used to the wires, and

A young Berkshire pig sniffs an electric line around its pen.

Young pigs need to get used to electric wires, and may rush at the fence, so expect some escapees at first.

often they will rush at the fence and push through, despite being given a shock. Then they stand on the other side of the fence and refuse to come back! So expect some escapees at first. Wires are more effective than tape, but they are not as visible, and the pigs do not see them easily. We have used a single strand of tape along with two wires, so the pigs are able to see a visible marker.

Food and water

There are plenty of commercial pig foods available, with different formulas for weaners, growers and sows, which vary in the protein levels. The usual recommendation for feeding weaners is to give ad-lib (unlimited) weaner pellets until they are about 10 weeks old. However, we have found that with the slower-growing traditional breeds this is best avoided. Our young pigs seem

This recycled plastic container has served us for many years, proving to be resilient to the pigs.

to do better on a lower protein level of about 15 per cent, so they do not grow too fast. Our preferred food is an organic sow roll, the largest size of pelleted food, as there is less waste.

The amount of food the young pigs get is increased gradually, based on 0.5kg (1lb) per month of age, up to a maximum of 2-2.5kg (4lb 6oz to 5lb 8oz) per day. Check that the pigs have eaten their food in about 20 minutes or so – if they haven't, you are giving them too much. Ideally, pigs are fed twice a day, so you can split the daily food ration.

On average, expect to use about 250-300kg (551-661lb) of food to get a weaner to slaughter weight, depending on the breed. Their food can be supplemented by roots and fodder from their pen, plus any windfalls, acorns and fruit from the hedgerows, etc. They will happily eat roots and leaves and other vegetables from the plot, but you must not feed them any catering waste, from your own or a commercial kitchen or anywhere that sells meat (see Appendix). Food is usually given in troughs, but during dry weather it can be scattered on the ground. This helps to prevent bullying. Sow rolls are particularly good for scattering, and the search for food keeps the pigs occupied.

Each pen needs a supply of clean water. The trough should be made from a sturdy material such as galvanized iron, recycled rubber or heavy plastic. Pigs are expert at making their water muddy, so the trough needs to be tipped out and refilled daily. In summer they will need a wallow where they can lie in the mud, which will help to protect their skin from sunburn (see opposite).

Health and welfare

Pigs are generally very healthy, and by observing your animals you will soon pick up any signs that something is not right. For example, they may have runny eyes, a discharge from the nose, or they may have diarrhoea or be lame.

Parasites

As with poultry, one of the most common problems with pigs is parasites, either ecto- or endoparasites. For example, they can get lice or suffer from a range of intestinal worms. Pig lice are large, so you can spot them moving around the head and ears, and they can be treated with a louse powder. Mange is caused by mites: it is an irritant which causes them to rub and scratch, and needs to be treated with ivermectin or a similar product, which you get from your vet.

Many people worm their pigs routinely, either with fenbendazole mixed into their food over a number of days, or with injectable ivermectin or its derivatives, such as doramectin and paramectin, usually starting at 8 weeks and then at 6-monthly intervals. When you buy your weaners, ask if they have been wormed. If they have and if you intend to slaughter at 6 to 7 months, you may not need to do it again, especially if you are housing them on fresh ground. Don't forget that there is a withdrawal period for these treatments, during which time the animal may not enter the food chain.

Being organic, we do not routinely worm our pigs, as we only keep them for a short time and we rotate the pens to reduce any worm burden. Rather than simply give the pigs a wormer, it's much better to do a faecal test first to see if there is any need. This is true for all livestock, not just pigs. The overuse of wormers is leading to resistance, so it's important to use the treatments only when necessary. It is far better to manage the land appropriately to reduce the incidence of worms.

Lameness

Lameness in young pigs is not uncommon. It can be temporary, as a result of an injury sustained during rough play or from running on hard ground in dry weather, in which case the problem resolves itself in a few days. Sometimes, the lameness is due to arthritis caused by bacterial infection, in which case you need to speak to your vet to determine the best course of treatment.

Sunburn and heat stroke

If you have a white-coated pig, sunburn is a consideration in summer. Believe it or not, you can use sun cream on your pigs, especially on the ears and snout. Watch out for heat stroke during periods of hot weather, and make sure there is plenty of water and a wallow. But even then, pigs can overheat easily. The symptoms of heat stroke include shallow, rapid breaths and trembling, with the pig generally looking very poorly. Don't tip cold water over it, as this can cause the blood vessels near the surface of the skin to constrict and cause even more problems. Get the pig into the wallow or under shade and if in doubt call your vet.

REMEMBER that if you do have to use medicines or call a vet to treat your animals, you must keep good records, recording what was used and how much, the date, batch number, the animals treated and the withdrawal period (see Appendix).

Make sure your pigs have access to a wallow in summer.

How much does your pig weigh?

It can be difficult to judge the weight of a pig, and often you may think it's not large enough to slaughter, while in reality it's plenty heavy. You can work out the weight of a pig using a piece of string! (A string is softer and easier to wrap around the middle of the pig than a tape measure.) Take your measurements while your pigs are eating and standing still! Then use this formula to calculate the weight, which is usually accurate to within a few kilos or pounds:

Weight (in kilos) = (G^2 x L [metres]) ÷ 69.3
Weight (in pounds) = (G^2 x L [inches]) ÷ 400

Where G = girth (the circumference behind the forelegs), and
L = length of pig from base of ears to base of tail, in either metres or inches.

Make a note of the girth reading. The abattoir will tell you how much your pigs actually weighed, so next time you can simply measure the girth and know if they are ready to go, or if you have to leave it a few weeks.

Slaughtering your pigs

A commercial pig raised in a barn may be ready to go to slaughter by 16 weeks, but traditional breeds raised outdoors take much longer. Middle Whites or Berkshires run to fat quite quickly, so are taken at about 60kg (132lb) at around 6 months; a pig of this size is referred to as a porker. A Gloucestershire Old Spots, Large Black, Tamworth or Saddleback will be ready by about 7 to 8 months. Those pigs destined for bacon and gammon can be run on for 8 to 10 months, when they can reach as much as 120kg (264lb), to give larger hams; these are commonly called baconers.

When the time comes to slaughter the pigs, make sure you have booked into your local abattoir well in advance. It pays to visit the abattoir beforehand to find out when and where you have to take your pigs and unload them, and to talk to them about the butchery services they

Don't feed your pigs the night before you move them, as it's easier to get hungry pigs into the trailer.

offer. Also make sure that all your pigs are suitably identified with either a metal tag or a slapmark (tattoo). If you are using metal tags, you may have to put in your order for tags well in advance. Also, all your paperwork needs to be in order (see Appendix for details of these rules).

You will also have to arrange transport to the abattoir, so if you don't have a trailer you may need to borrow one from a friend or neighbour. It's important to get your pigs used to the trailer, so that they are not stressed by being loaded on to it. You can do this by backing the trailer into their pen and feeding them it in for a few days. Do not feed your pigs the night before you intend to move them, as it's easier getting hungry pigs into the trailer! Once they are in the trailer, you can tag them and make sure they are clean enough, as abattoirs can refuse muddy pigs. If they are dirty, clean them off with a bucket of water and sponge. Once the pigs are delivered, the trailer must be washed down and disinfected. This can be done at the abattoir or when you get home.

Most abattoirs have a cutting room where they can butcher your pigs. If not, you can arrange for the carcasses to be taken to a local butcher. In addition to joints, chops and mince, you can order some sausages and arrange for some meat to be cured for bacon and gammon. You could opt to butcher your own pigs at home, but if you do so there are restrictions on selling the meat.

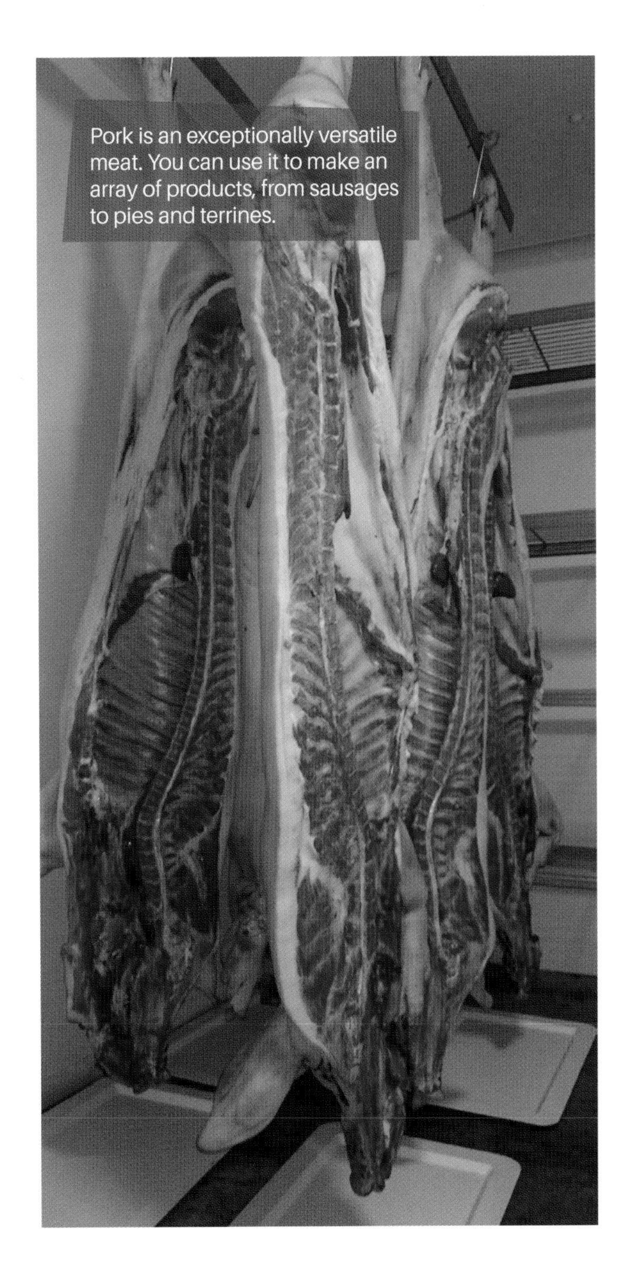

Pork is an exceptionally versatile meat. You can use it to make an array of products, from sausages to pies and terrines.

chapter NINE

Including sheep or goats

If you'd like to keep sheep or goats on a plot of around an acre, you will need a decent area of permanent grass. Then you could raise a few lambs for meat over the summer. Goats need more shelter but less space than sheep, so are an option for keeping year-round.

A Southdown ewe and twin lambs. This is a small, easy-to-keep breed, good for the novice smallholder.

Keeping sheep and goats on a small plot is a challenge, and more so if you decide to keep them throughout the year. Sheep are grazing animals, so the demand on the grass will be high, especially in winter, and you may need to rely on imported fodder or sheep nuts to ensure they have enough food. Goats are browsers, and they nibble trees and shrubs as well as grass. They may need imported fodder in winter as well.

On a plot approaching an acre in size you have a few options. The bigger the area of grass you have, the better – so aim for at least half the plot. Anything smaller will not be enough for sheep or even goats. One approach is to raise some lambs over the summer months, when the grass is at its most productive, and let the grass rest for the remainder of the year. Alternatively, you could run a few animals under an orchard so you can make dual use of the land, but remember that shading from the trees and competition for nutrients will reduce the quality of the sward, and the trees will need protection.

Understanding grass

If you plan to have sheep, keeping a quality sward will be critical, so it pays to invest in your grass, as it is your cheapest source of feed. Your permanent pasture probably consists of several

It's important not to grow too much grass for the number of sheep, since sheep like to nibble short grass.

different grasses, of which the most common is perennial rye grass (*Lolium perenne*), which is a hard-wearing species. Grass growth is at its fastest in late spring, when the rye grass is producing a new leaf every 4 or 5 days, but in the middle of winter this can take 30 days.

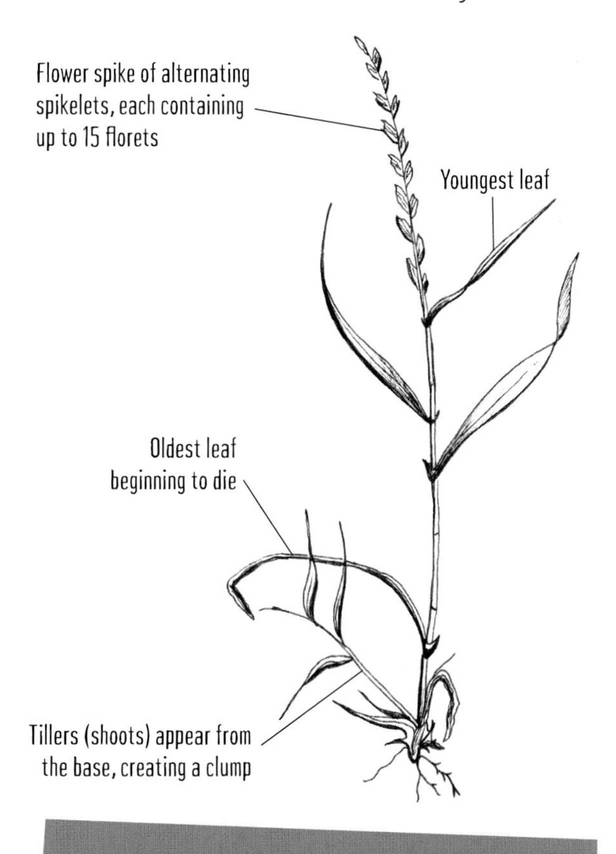

Perennial rye grass has three leaves, and as the fourth leaf starts to grow, the oldest one dies. It produces tillers (side shoots), which create a clump.

Grass is influenced by many factors, such as soil temperature, water and nutrient availability, and light levels. In spring, when grass is growing at its fastest rate, a hectare (2½ acres) can gain 20-60kg (44-132lb) of dry matter in a day. But if the grass is very short it will never reach this level of production, because the small leaf surface area restricts photosynthesis. Production is slower with long grass too, as the long leaves shade each other and some of the leaves are dying, so the maximum rates are only achieved on swards of medium height, where there is large leaf surface and less leaf death.

If possible, you should aim to graze a sward in optimum growth and to stop grazing when the grass height is too short to sustain production. For grazing sheep, it's important not to grow too much grass for the number of animals, since sheep like to nibble short grass. The best-quality grass for sheep is when rye grass is approaching the three-leaf stage, so you need to prevent it getting to the fourth-leaf stage, when leaves start dying (see diagram, left). This leads to wastage and the build-up of dead matter around the base of the plants, which smothers adjacent grass plants. When the grass is at the optimum stage, the animals will be able to make use of about 70 per cent or more of it, but if the grasses are long and tough, the animals may use less than half of what is available grass. The same is true if the sward contains weeds such as thistles and nettles.

Improving your grass

Many paddocks are sown with the hard-wearing rye grass, but you would be better to opt for one that is a rye-grass-and-legume mix, as legumes offer your animals a protein-rich diet. Legumes, such as clover, are beneficial plants. Not only do they fix nitrogen and so boost the productivity of the surrounding grasses, but also they are favoured by sheep over grass. The rate at which lambs gain weight is much higher on clover, in some cases by 25 per cent more, and furthermore the clover tends to continue growing through the summer, when grass production dips. If there is little clover in your pasture, you could improve the situation by oversowing it with red or white clover seeds in spring or early autumn (see opposite). Ideally, you should aim for 40-60 per cent clover in mid summer.

You could also include medicinal plants, such as chicory (*Cichorium intybus*) and sheep's parsley (*Petroselinum sativum*), which are known to have anti-worm properties. Some seed merchants now sell a herbal grazing ley mix, which is a mix of grasses with, for example, clovers, trefoil, lucerne (*Medicago sativa*, also called alfalfa), chicory, salad burnet and yarrow, and can be bought in quantities suitable for small paddocks. The deep-rooted plants, such as chicory and lucerne, are good for drawing up nutrients and water and improving the soil structure.

The welly test

It's easy to assess your grass length. Simply walk into the paddock wearing wellington boots. With just 4-5cm (1½-2") of grass, the sward barely reaches the toes of your boots. Ideally, the grass needs to be 8-10cm (3-4") – reaching above your ankles, with the tops of the leaves bending over. If it's halfway up your boots, it's getting too long for sheep.

You should aim to have about 8-10cm (3-4") of grass before you introduce the sheep, and let them graze it down to about 5cm (2"), then move them to the next paddock. If you can't rotate the areas, don't let the grass get shorter than 6-8cm (2½-3") – if it does, you are grazing too many animals.

The low growth habit of white clover is ideal for grazing sheep.

A herbal ley mix with salad burnet.

Each spring, try to harrow the grass (drag it with tines) to remove dead leaves and molehills, and if necessary oversow it with clover. This literally means scattering seed over the existing sward, with no need to plough up the existing vegetation. Do this when there is plenty of moisture in the ground. The recommended rate is 5kg/ha (4lb 6oz / acre) of seed. Then graze the sward to remove some of the existing grass and allow light through, so the seedlings can get established. On small paddocks, an application of calcified seaweed can be beneficial to replace depleted nutrients.

Mob grazing

To maximize the productivity of your grass, try a 'mob grazing' approach. This is a style of grazing that mimics the way herds of wild grazing animals move across a grassland. They graze an area intensively and then move on to fresh pasture. While they are grazing, they trample much of the grass, creating a thick mat that protects the soil underneath from compaction and erosion and provides the soil microorganisms with food, together with the animals' dung and urine. This all helps to increase the organic matter in the soil and boost its water-holding capacity.

This is an effective way of grazing, even in a small area. The key is to have multiple pens or paddocks, so you can rotate your animals between them. Although this is costly to set up, owing to the extra fencing required, it means more productive grass. With this method, your animals get to graze fresh grass at its best. They are then moved on before they overgraze it and damage the grass plants. It also reduces the risk of poaching (sodden and trampled ground) in wet periods, which is important when you only have a small area of grass to sustain. Another benefit of rotation is the reduction in the worm load in any given area. If you keep ducks and geese, you could include them in the rotation. Geese prefer shorter grass, so they can follow after sheep, and both ducks and geese will help to clear the land of parasitic worm eggs. (See Chapter 7, pages 142-3 and 158.)

Keeping sheep

You may be thinking of a couple of lambs for meat, or perhaps you simply want to have sheep to keep control of the grass. Either way, you have decisions to make about the type and number of sheep that you can accommodate.

Once you understand grass management, you can work out how many sheep you can afford to graze on your grass without having to resort to extra fodder. The recommendation is between three and five ewes per acre (seven to twelve per hectare) on good rye grass and clover leys. So if you are considering buying in weaned lambs and growing them on to slaughter weight, you could probably accommodate three to five lambs on half an acre, as you would be keeping them during the most productive months of the year. But be aware that if there is a summer drought, grass production could halt and you would have to buy in fodder as silage or hay.

Ideally, you should aim to buy weaned lambs of about 8 to 12 weeks old in spring – before the grass has had a chance to get away and become too long – and have them slaughtered in early autumn. This means looking for farmers who lamb early in the year. If you can't get the lambs until summer, you can top the grass to keep it from getting too long.

Lambs usually go to slaughter when they are about 20 weeks old and weigh around 40kg (88lb), although this does vary with breed. If the lambs have not put on enough weight by late summer, you can feed a high-quality concentrate to bring them up to weight. The lamb carcass that you get back from slaughter weighs about half that of the live lamb.

Dorset Downs can be tupped (mated) early, so lambing occurs in late winter and the lambs are well grown in spring.

The sheep and lambs will need secure fencing, some form of shelter from the sun if there is no natural shade from trees, and a water trough. When planning your paddocks, it can help to position the water centrally so you can rotate the animals from paddock to paddock and they can still access the water.

Essential equipment

You still need a fair amount of equipment, even for just a couple of lambs, but it is an investment as it can be used for a number of years. Top of the list are sheep hurdles: small movable fences that you can use to contain the sheep. Handling the animals and getting them used to being rounded up and contained within a ring of hurdles makes management so much easier, especially when you want to shear them or get them on to transport for slaughter. You will need a pair of hand shears for dagging (that's clipping mucky fleece around and under the tail), plus a pair of hoof trimmers to trim the hooves back when they get too long. Other bits and pieces include buckets, feed trough, antiseptic spray and foot spray.

Routine care

Sheep are demanding in their care. They need to be checked twice daily. Surprising though it may seem, sheep can roll on their backs, get stuck and die! Every couple of months they have to be dagged and their feet checked for foot rot, and in late spring they need to be sheared. A few weeks after shearing they are sprayed to prevent flystrike, caused by parasitic flies – a major problem in lowland sheep, which can kill if left untreated. They may also need vaccinating against Pasteurellosis and some of the clostridial diseases.

The smallest sheep in the world

If you are determined to have sheep all year round, then one option is to keep a couple of ewes of a small breed, such as the French Ouessant. These are small, friendly sheep that are happy being handled, and are gentle and easy for a first-time sheep owner. As adults they weigh just 20kg (44lb) and stand 45-50cm (18-20") tall. The rams have impressive curled horns. Since sheep are social creatures, you would need to keep a minimum of two animals. They are suited to small paddocks and orchards, although any trees will need protection.

Some of the primitive breeds are also small, the Soay for example – but they are quite adept at escaping from paddocks, unlike the Ouessant.

Ouessants come in three colours – black, brown and white – so their fleeces are popular with spinners.

orphan lambs

Another option is orphan lambs. These are lambs that may have been abandoned by the ewe, or the ewe has died, or there are triplets or quads and the farmer wants to reduce the pressure on the ewe. Orphan lambs are normally just a few days old and have to be bottle-fed with a milk replacement. This can be time-consuming and the lambs can be sickly, but if they are healthy you will get some friendly lambs for your paddock. They will get very attached to you, so if you go down this road it can be better to buy ram lambs, as there is less temptation to keep them and they reach slaughter weight more quickly.

Many people think goats will eat anything, but in reality they are fussy eaters, far more so than sheep.

Sheep are usually wormed regularly, and treated for liver fluke if it is prevalent in the area. If you are keeping your sheep year-round on small paddocks, it is highly likely that you will get a worm problem caused by a build-up of worm eggs on the ground. But before you buy an expensive bottle of wormer, send a faecal sample to your vet or worm lab for a worm count to make sure. This avoids unnecessary worming of animals, which is contributing to the growing problem of worm resistance around the world. Regular faecal testing of your sheep will allow you to keep on top of any problem.

Keeping a couple of sheep can be expensive, as all the wormers and other medications come in large bottles, aimed at commercial farmers, while vaccines are in 20-doses. Plus, unless you can shear your own sheep you will need the services of a shearer, who usually charges a set-up fee plus a small fee per animal. Think carefully before committing to keeping sheep, as they can be a burden, unlike pigs, which are much easier to care for.

Keeping goats

Goats are native to much warmer climes than that of the UK, and they are not hardy like sheep, as they lack the thick, waterproof fleece. They don't need as large a paddock as sheep, but they do require a shelter in the form of a small stable or pole barn, as they cannot be out in all weathers. If you have stables or other buildings on your plot, then goats may be an option.

Shelter and pasture

Goats need a goat house in which they can be kept during cold or wet weather, such as a barn, stable or a large shed. It should provide at least

Three goats are kept on this small paddock with a shelter. The trunk of the tree, which provides shade, is protected from the goats.

3m² (32 sq ft) of floor space per goat and be tall enough for them to stand on their hind legs, with a layer of bedding plus bales for them to stand on. If you house them in a barn and move them to pasture daily, you will also need to have a field shelter.

Some people keep their goats in a good-sized yarded area all year round; others for just the colder winter months. If the goats are kept in a yard, they will need a daily supply of hay, hedgerow clippings, etc., as they require a varied and fibrous diet, plus fresh water. Although housing the goats this way allows you to keep a small herd in a restricted area, they do need enough space to exercise freely and they also need an enriched environment.

Others house their goats at night and turn them out on to a well-fenced paddock to range free each day. Although goats don't need as large a ranging area as sheep, they do need a good-quality pasture and ideally one with access to a hedgerow. A half-acre paddock could probably accommodate between three and four goats, depending on the breed. As with sheep, too many animals can lead to overgrazing and soiling of the grass.

Many people think that goats will eat everything and anything, but in reality they are fussy eaters and are far more selective than sheep, preferring to browse rather than graze, so less of the grass may be used. You could run goats with geese to help keep the grass down, otherwise you may need to top the grass during spring and early summer. If goats cannot access their night shelter during the day, they will need a field shelter, as the lack of lanolin in their hair means they cannot tolerate rain.

Expect goats to eat any foliage in reach of their long legs and necks.

Goats are known to be escape artists. They like to clamber to the top of things, so fencing needs to be secure, and higher than sheep fencing, ideally 1.2m (4'). Otherwise they will soon break out and defoliate a nearby orchard or veg patch.

The choice of goat

There are three main groups of goat: dairy (Nubian, Toggenburg, Saanen, Alpine, Golden Guernsey), meat (Boer) and wool (Angora and Cashmere), plus the pygmies. Like the sheep, the goat is a social animal, so you need at least two goats, although they don't have to be of the same kind (you could opt for a dairy goat and a companion pygmy goat, for example). Pygmy goats are very popular because of their small size, and are often seen on petting farms.

The best place to buy a goat is from an experienced breeder, who will be able to advise you on care. Females are more expensive than males. Often people buy a goatling, a young goat that they plan to get into kid, but animals of 4-5 years old are cheaper and probably easier to handle. Don't be tempted to buy an uncastrated billy goat unless you are breeding, as they smell and can be difficult to handle, although castrated billies can make good companion animals. You should check that any animals you buy have been blood-tested for Caprine Arthritis Encephalitis Virus (CAEV), which is an incurable disease.

Essential equipment

Goats drink a lot of water, especially in the case of lactating nannies, which take up to 20 litres (approx. 4 gallons / 5 US gallons) a day, and they are fussy about their water – it must be clean, so a constant supply of fresh water is essential. Other essentials include a feed trough that is off the ground, hay net, buckets and bucket holders, plus basic veterinary equipment (similar to sheep, see page 181). You will also need somewhere dry to store hay and straw. The soiled bedding in the goat house must be replaced regularly, so you will need a dung heap on site.

Feeding your goats

As mentioned already, goats are browsers rather than grazers. They tend to search upwards for food, standing on their hind legs to reach hedgerow leaves and branches. As they can be fussy eaters, they may not eat the hay provided! Their diet can be supplemented with goat nuts. If you are feeding hay and nuts you will need to supply 2-3kg (4lb 6oz to 6lb 10oz) of a good-quality hay and up to 1kg (2lb 3oz) of goat nuts daily, depending on the size of the animals. Always supply

These Nubian goats are fed a good-quality organic hay every day.

Goats are escape artists, and like to clamber to the top of things. They stand on their hind legs to search for food.

the hay in a net, as anything falling to the ground will not be eaten. If you are bringing in forage such as green food from hedgerows, allow at least 7kg (15lb 7oz) because of the wastage and water content. Bear in mind that young goats may eat any poisonous plants that have been inadvertently included, so you must be aware of which plants not to fed to a goat; also, plants that are safe for sheep may not be safe for goats (the list is surprisingly long and varies around the world, so check details on local goat society websites). A mineral lick can be useful to ensure that they get all the essential nutrients, especially if you are feeding a restricted diet.

Basic care

Goats are not quite so demanding as sheep, but they still need checking twice daily. The routine care is similar to that for sheep (see page 181) – foot trimming, worming, and vaccination against Pasteurellosis and some of the clostridial diseases. Angora and Cashmere goats will also need shearing; Angoras twice a year.

The Golden Guernsey is a popular choice for beginners.

Goats' milk

Many people are turning to goats' milk for its nutritional benefits. Compared with cows' milk, it contains less lactose, while it has more fatty acids, calcium, potassium and vitamins A, B5 and C. It is also more nutritionally dense, so you do not have to drink so much of it to get the benefits. Its composition is closer to human milk than cows' milk, so it is easier to digest, and people with allergies to cows' milk can often tolerate goats' milk as it has a different protein.

The dairy goat

A dairy goat requires much more commitment in time than other types, as she will need to be milked twice a day during her lactation. The main dairy breeds are Nubian, Alpine, Saanen, Toggenburg and Golden Guernsey. The highest yielding are the Saanen, producing as much as 1,200 litres (263 gallons / 317 US gallons) per lactation, while the Nubian produces milk with a higher fat content. The females are ready to be mated at 10 months and their gestation period is around 5 months. Most give birth to two kids. You will need to keep a billy goat too, to get the females in kid. The average productive life for a dairy goat is 7 years.

If you are considering a small herd of dairy goats, then one of the best options for beginners is the Golden Guernsey – a friendly, docile breed that is smaller than other dairy breeds. They can be expected to yield 2-4 litres (3½-7 pints / 4¼-8½ US pints) of milk a day, which is a lower yield than from other breeds, but they also require less feed and have a better feed-to-milk conversion rate. The fat and protein content of the milk is perfect for making cheese and yoghurt. Once a goat has had her kids and they have been either removed or weaned (anything from 3 weeks to 3 months), she can be milked for up to 9 months, although the milk yield declines towards the end of this period.

Equipment-wise, you will need a milking stand and a clean area where you can milk, plus a stainless-steel milking bucket, udder cleaning wipes, milk filters and bags, as well as equipment and a suitable area for cheese-making. Once collected, the milk needs to be stored somewhere cool, where it is bagged up. With a small herd, milking is done by hand, so it can be quite time-consuming and of course has to be done every 12 hours.

Unless kept for breeding, most billy kids are raised for meat.

The meat goat

Goat meat is rising in popularity. It's a healthy, lean meat, with a flavour more like game than lamb, but is still quite expensive, so raising a few kid goats for meat may be an option. The Boer is the traditional meat breed, as it has been bred with meat production in mind, with a good-sized frame and well-developed muscles.

In many commercial dairy herds, male kids have no value and are often killed at birth. But this is changing as the market for goat meat increases, and increasingly they are sold to be raised for meat. Indeed, Boers are often crossed to a dairy breed to get better-quality male kids for meat. The kids are nursed for around 5 days, as this first milk from the new mother is not used by the dairy because of the high colostrum content, and then the dairy can earn some money from selling the kids.

Raising goat kids is time-intensive at the start, as the young kids are fed an artificial milk replacement until they are weaned. They are then moved to a diet of straw, hay and goat mix. They are raised for about 6 months, until they have reached around 35kg (77lb) in weight and are ready to be slaughtered. The finished carcass weighs around 15kg (33lb). The goats will have to be slaughtered at an abattoir, so check that your local one will take them.

chapter TEN

Aquaponics

Aquaponics is a fascinating, simple and highly productive system that combines fish and plant crops. The dirty water from the fish tanks flows through grow beds, where bacteria convert the nutrients to a form that the plants can use. It doesn't need much space, and you can even build your own DIY set-up.

Many years ago while on holiday in Kenya, I visited an innovative project based in an old cement quarry that was being restored by Dr Rene Haller, a Swiss environmentalist. The restoration work turned the derelict quarry from barren waste to a rich forest ecosystem, part of which is now a wildlife reserve. Alongside the restoration were numerous smaller projects to develop alternative technologies and promote organic farming methods. One of the projects was keeping fish.

They had built a series of fish ponds, with water draining from one to the next. The ponds were stocked with tilapia, and the dirty water drained to the lower ponds where the locals were growing Nile cabbage and rice. The water draining in turn from the vegetable plots was circulated back up the slope to the fish ponds. The plants took up the nitrogen from the water before it was circulated back to the fish. Surprisingly little water was lost, so this precious resource was recycled many times over. It was a very simple system, as many of these best ideas are, using appropriate technology. The fast-growing tilapia fish provided the locals with a valuable, but cheap, protein source plus a supply of rice and vegetables. Over the years the design has been improved and now the system includes chickens, whose droppings are fed to the fish! Today, we call such a system aquaponics.

Aquaponics is attracting a lot of interest, and it is now much easier to get hold of both fish and equipment, and even to buy kits. It is a technical

An overhead view of the growing tanks and rearing pens at Baobab Farm in Kenya.

A commercial system at Growing Power's urban farm in Milwaukee, USA, which produces tilapia and yellow perch, watercress and tomatoes. The water is drawn through one pump and gravity-fed through the grow beds and back to the fish tanks.

topic, so in this chapter I am offering just an introduction to the subject. If this overview sparks an interest, you will be able to find a wealth of more detailed information online and in books (see Resources).

Why aquaponics?

People have kept fish for thousands of years, both in ponds and in netted cages in shallow coastal waters. The disadvantage of rearing a large number of fish in ponds is the large amount of water polluted with fish waste that has to be discharged and replaced, so it is not surprising that growers looked to combining hydroponics – growing crop plants in soil-less culture – with aquaculture. Commercial aquaponic systems were established in both North America and Australia during the 1980s, soon followed by the development of smaller systems suitable for the garden or smallholding.

Aquaponics is a simple system comprising fish, plants and bacteria. The fish are kept in large tanks and their dirty water, laden with ammonia and other wastes, flows through grow beds, where nitrifying bacteria convert the ammonia first to nitrite and then to nitrates, a form of nitrogen that can be taken up by plant roots. The water is effectively filtered as it passes through the beds, and clean water is returned to the fish.

Plants in an aquaponic system grow up to three times faster than in soil, without additional fertilizers.

The circulation of water from fish to plants and back again can take place over and over again, with the only losses being from evaporation and transpiration from the leaves of the plants. It is a highly productive system, with plants growing up to three times faster than in soil and without the need for additional fertilizers. A small system has the potential over a 6-month period to yield about 25kg (55lb) of fish and many kilogrammes of vegetables from beds just 3m² (32 sq ft).

The options

There are three main types of system suitable for a small-scale aquaponic system:

A **media-based system** is probably the most common, in which the plants are grown in beds of inert planting media, such as clay pellets or gravel. The bacteria form a biofilm over all the surfaces, both inert and living, that are in contact with water, and they convert ammonia in the water to nitrates. The planting media also filter the solid waste. The grow beds can be used to raise a wide range of crop plants, from salad leaves to aubergines, tomatoes and exotics.

The **raft system** is often the choice for people wanting to grow salads and other fast-growing crops. The waste water is filtered of solid waste and then pumped into a shallow channel, on which floats a foam raft. The plants are planted into the raft so that their roots dangle in the water, from which they take up nutrients. Alternatively, the raft is suspended in a tank filled with circulating water from the fish tank.

The **hybrid system** is a cross between the media-based and the raft. The waste water is pumped first on to media beds, which remove the solid wastes, before being emptied into a raft system.

The media-based system is the most straightforward, and you can now buy complete kits (minus the fish) online. However, these can be quite expensive, and it is possible to make your own DIY system, which I describe on page 195.

A simple media-based system

This system consists of a large tank, which houses the fish, a grow bed filled with planting media, a pump and lengths of pipework. More tanks and grow beds can be included. The tank size determines the number of fish that you can raise, with a realistic stocking rate of 3kg of fish per 100 litres (6lb 10oz per 22 gallons / 26 US gallons). In turn, the number of fish determines the area of grow bed you can support (see page 198). The grow bed contains about 25-30cm (10-12") of media, such as gravel or clay pellets of a particle size around 8-15mm (¼-⅝"). The particles support the plant roots and provide a large surface area for colonization by nitrifying bacteria. The resulting cleaned water is returned to the fish tank. Regulation of the nitrogen levels throughout the system is critical to its smooth

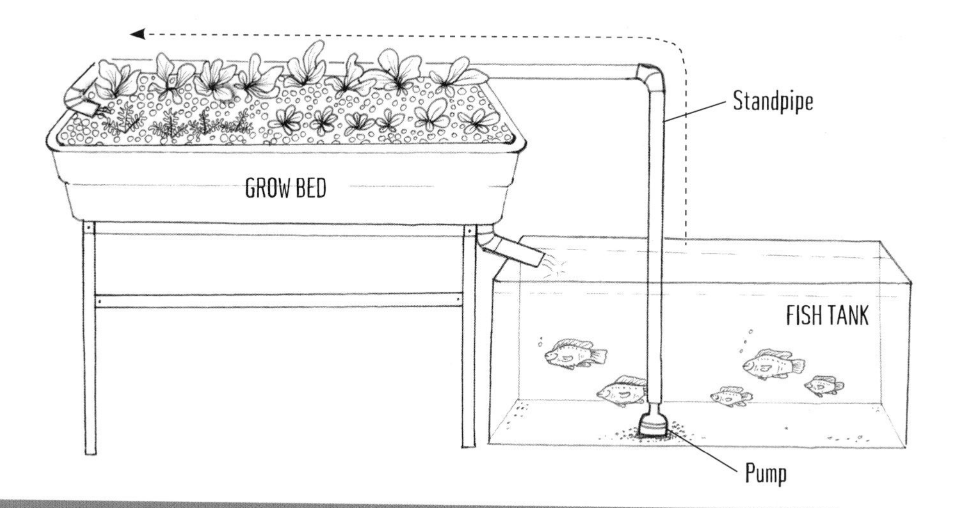

A simple media-based system with flood-and-drain circulation. The water level in the fish tank falls as water is pumped out on to the grow bed, and rises again as water drains back in.

functioning, but, once the system is established and working well, it can run for months with just the occasional tweak.

The water is circulated in a number of ways. The simplest system is called flood and drain, whereby the grow bed is located higher than the fish tank, and a pump in the fish tank pushes the water up a standpipe into the grow bed, where it drains through the media and returns to the tank. With this arrangement, the water level in the fish tank varies (see diagram above). If the pump is located in a second tank, known as the sump tank, the water in the fish tank is maintained at a constant level (see diagram on page 196).

The sump tank

A sump tank is needed as soon as the volume of the grow bed or beds exceeds the volume of the fish tank. The run-off from the grow bed accumulates in the sump tank, so it has to be positioned lower than the grow bed. It allows the water in the fish tank to be maintained at a constant level, which is better for the fish and means more water can be added to the system, which helps to maintain stability.

A sump tank works on this basis: at the start, the fish tank and sump are full of water and the grow bed is empty. When the pump starts working, water is moved from the sump to the fish tank, raising its level and letting water flow out into the grow bed. The grow bed fills with water and starts to drain out into the sump. Once the grow bed is full, the pump stops and the water finishes draining back out into the sump. In some systems, the pump is on all the time, keeping the grow bed in constant flood.

Temperature considerations

In a climate such as that of northern Europe, you have to decide whether you run the system year-round or just during the warmer months. If you do wish to run it all year, you will need heating for the fish tanks and also additional lighting for your media beds in order to maintain plant growth through the shorter days of winter. Many

A tank of tilapia that is part of a media-based system.

people setting up a system for the first time choose to grow carp or perch, either in a polytunnel, greenhouse or outside, and run the system from spring to autumn. The fish survive over winter, but as the temperatures are too low they do not grow, and need little food or oxygen; then they pick up again when the weather gets warmer in spring.

If you decide on a year-round system, choose carp or tilapia. Tilapia need a constant 25-30°C (77-86°F), so in temperate climates their water will need to be heated so they can survive and grow all year. The heating cost is the second largest expense after the fish food, but can be reduced by careful design of the structure in which the fish are kept. You could use an insulated polytunnel or adapt a brick building, for example. You could use photovoltaic panels to heat the water, with an electric water heater as a back-up. Don't forget that your electrical system will be required to power not only the pump but also the lighting, and probably to meet at least some of the heating demand.

Choosing your fish

There are a number of options when it comes to the fish, but some are better suited to a small-scale system and are more resilient.

Tilapia

If you are setting up a system for the first time, you would be wise to opt for an easy fish such as tilapia, but, as already mentioned, they do require heated water. Tilapia is the most widely raised fish in aquaponics systems because it grows fast, is relatively tolerant of water conditions, and the fingerlings (young fish) are easy to source, as is their food. The fish reach a marketable size of around 500g (1lb) in just 6 months, so it's possible to stock a system in spring and harvest in autumn.

Carp

This is an easy fish to grow, with a fast growth rate, but it's not to everybody's taste. The optimal temperature to raise carp is 20-24°C (68-75°F), but they do not need heating. As temperatures fall, so does their growth rate, so they can take a year or sometimes longer to reach a harvestable size. They can be grown in outdoor pools too (see page 199).

Rainbow trout

A fast-growing fish, the rainbow trout requires a high-quality protein diet and good-quality water. The ideal water temperature is 14-16°C (57-61°F), but they will survive at lower temperatures. However, their growth rate falls both above and below this temperature range. The fish take up to 10 months to reach size.

Perch

This species has a relatively fast growth rate and needs a high-quality protein diet. The optimum temperature range is 16-20°C (61-68°F), but heating is not essential. Although perch is an easy fish to keep, it's not to everybody's taste.

Setting up a DIY tilapia system

By far the easiest system to build yourself is one based on the ubiquitous IBC (intermediate bulk container), each of which measures a cubic metre (35 cubic feet), with a capacity of 1,000 litres (220 gallons / 264 US gallons). They come with a supporting metal frame. They are found worldwide and can be picked up cheaply second-hand. A small system based on a fish tank made from one IBC would support 3m^2 (32 sq ft) of grow beds. To make the fish tank, the top of the IBC is simply cut off for access, while two other IBCs are cut down to create the 3 x metre-square (3 x 11 sq ft) grow beds and the sump tank. As well as the IBCs you will need a water pump, air pump and air stone (see page 198), grow-bed supports (which could be concrete blocks and planks), growing media such as volcanic gravel or expanded clay, plumbing materials including pipes and connections, plus bits and bobs such as a fish net, weighing scales, tape measure (to measure fish growth), thermometer, shading materials and a water-testing kit. Also, depending on your location, you will need an electric or a photovoltaic-powered water heater.

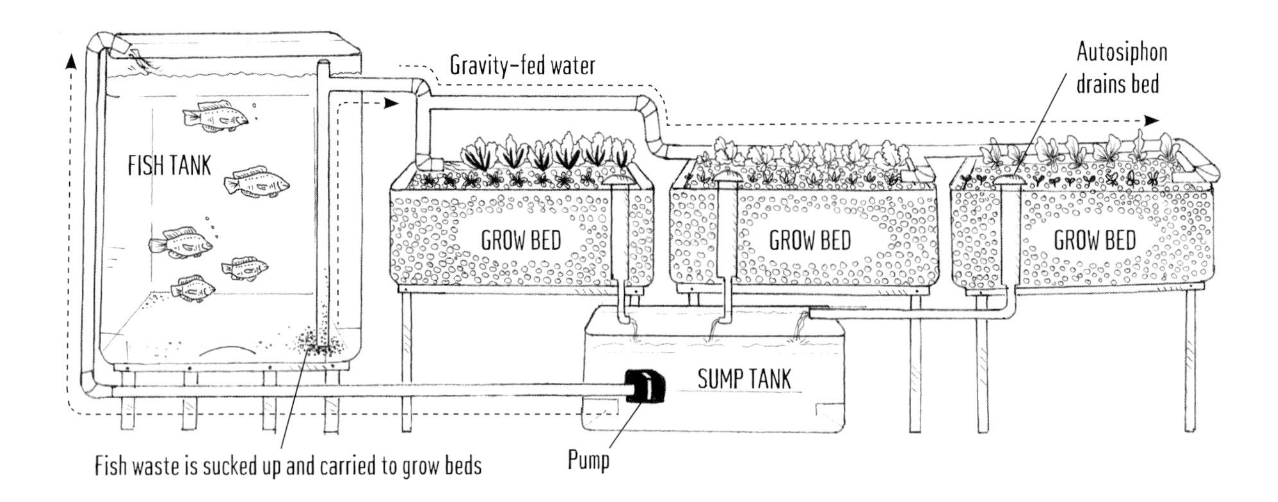

The layout of a DIY three-bed system made from three IBCs and using a CHIFTPIST ('constant height in fish tank pump in sump tank') system.

The IBCs last longer if they are coated with a dark paint to reduce damage from UV light. Also, this means there is less light reaching the water to stimulate algal growth, and the fish are kept in more natural conditions. Better still, box the fish tank to keep direct sunlight off and provide a more constant water temperature. You could add some floating plants to create shade and hiding places.

Check the source of your IBC

CHECK THE LABEL on the IBC to see what was transported in it and make sure the contents won't harm your fish, as some toxic chemicals can contaminate the plastic. Even if the contents were safe, you must wash the tank thoroughly, and it's easiest to do this out of the frame.

Water circulation

There are various ways of circulating the water, and the one suggested for this system is called CHIFTPIST (constant height in fish tank pump in sump tank). First, the water is pumped from the sump tank into the fish tank. To keep the water level in the fish tank constant, an overflow pipe carries the water to the grow beds. This pipe is called a SLO (solids lift overflow), and it extends to the base of the fish tank so it will draw up any solid wastes that have dropped to the bottom. The grow beds fill with water, and when they are full the autosiphon empties the water into the sump tank and the beds start filling with water again.

Typically, there is a flood and drain every 15 to 30 minutes, keeping the grow beds oxygenated for the plant roots and bacteria.

As soon as the plants are of harvestable size they should be replaced, to make best use of the nutrient supply.

The grow beds

These are made by cutting down two IBCs to form three beds of 35cm (14") depth (two beds come from one IBC – one from the top; another from the bottom). They are filled with growing media to within 5-10cm (2-4") of the top. It's best to use 8-15mm (¼-⅝") particles. If the particles are too small, there will be too few air spaces, and these are essential for oxygen to reach the roots and bacteria. I prefer to use expanded clay, as the particles are lightweight and create a large surface area. They will not interfere with the water pH, but they are expensive to buy. Crushed rock or gravel is cheaper, but is also much heavier, and you also need to check that it does not interfere with the water pH, which can happen if using crushed limestone.

You need to get the beds planted up quickly, as the whole system depends on healthy plant roots taking up the nutrients. Using transplants is a quick and reliable method. Wash off the potting compost from the plants' roots, so you don't contaminate your system. These will be productive beds, so don't waste the space: use closer planting distances than you would in vegetable beds. Aim to grow successionally too – as soon as the plants are of harvestable size they should be removed and replaced, to make good use of the continuous nutrient supply. Productivity can be high: for example, expect to be able to harvest 20 heads of lettuce or 3kg (6lb 10oz) of tomatoes per square metre (11 sq ft) per month in summer.

Where to locate your system

Your plants need sun, but your fish don't! So the location must suit both grow beds and fish tank, with extra-long connecting pipework if necessary. If you are housing the set-up in a polytunnel or greenhouse you may need to insulate the tank against winter weather and shade it in summer. Remember that you will need electricity and water supplies, and also easy access for bringing in the tanks, grow beds and media.

Testing the system

Once the beds are planted up and the pumps working, let everything get settled and the nitrogen cycling, all without the fish. This allows the beneficial bacteria to increase. Some people seed their system using a pond or aquarium filter from an existing system, as it will be full of bacteria. If you can't kick-start it this way, you just need to allow more time. You also need a nitrogen source for your bacteria. This can be achieved by adding urine, or a small amount of bleach, or some fish food to the tank, which gets pumped on to the grow beds for the bacteria to feed on. You could also add a few test fingerlings (young fish) or even goldfish. You can then use your water-testing kit to check the nutrient levels to see if everything is cycling correctly. If you are adding fish food, keep the nitrogen levels low, as it is so easy to create an algal bloom (an explosion in algal growth). Allow at least 6 to 8 weeks for the system to get established, monitoring all the time. During this time there may be fluctuations in ammonia and nitrite levels, but once the system is established they will stabilize.

During the test period the plants will be getting established and putting out roots. If they are not getting enough nitrogen, you will see signs of stress, for example leaves yellowing and slow growth. You can add a natural fertilizer such as a seaweed-based product to the water, which will boost the plants but not harm the fish. It may make the water black, but this soon clears up.

How many fish?

The number of fish you can support depends on the size of the fish tank and the area of the grow beds. A basic rule of thumb is that for each 500 litres (110 gallons / 132 US gallons) of grow-bed capacity you can support 25 to 30 fish, growing from fingerling size to around 500g (1lb) each. But you do have to consider other factors, such as oxygen supply in the tank on hot days, water flow, how many plants you are growing, the temperature, feeding regime, etc. Your IBC system of 3m^2 (32 sq ft) of grow beds (assuming 30cm/12"deep) will provide you with 900 litres (198 gallons / 238 US gallons) of grow bed, so in theory you could stock around 45 fish, with a harvestable weight of up to 22kg (48lb).

Similarly, you need enough fish waste to support the plants. With only one tank, you can have just one group of fish of the same age, so you need about 70 fingerlings at 50g (2oz) each in order to generate enough waste to feed the 3m^2 of plants. As they get larger, you can start to harvest from 3 to 4 months, when they are around 150g (5oz), to leave space for the remaining growing fish. Order your fish early to make sure they are available on time. Fish breeders usually transport the fish in plastic bags.

Fish care

The fish need a quality food to reach harvestable weight in 6 months, so a commercial food is probably easiest, at least when you are starting out. Fish are prone to a number of diseases, such as white spot infections, so daily checks are essential to catch problems early. Check the water levels and quality, and top up little and often. Don't pour in a load of chlorinated tap water! Fill a large, clean container with tap water and let it sit for a couple of days to allow the chlorine to dissipate. Remember that your system relies on electricity to power the pumps, so a power or equipment failure can quickly lead to oxygen starvation and dead fish. It is wise to have an alarm system and a back-up pump.

The water needs to be well oxygenated at all times. It picks up oxygen as it moves through the grow beds, but most systems also have an air pump fitted to an air stone at the bottom of the tank, which creates a stream of bubbles.

Check the rules

BEFORE YOU BUY YOUR FISH, make sure you have the right permits. Most countries have rules regarding the raising of fish, to prevent the spread of disease and parasites to native stocks, and which prohibit the release of farmed fish into waterways. In the UK, anybody raising fish for the table, from a small system in a garden to a large commercial system, must register with the Centre for Environment, Fisheries and Aquaculture Science (CEFAS - see Appendix).

Duck–fish farming

If an aquaponics system is too complicated for you, then you may want to consider duck–fish farming. This is not a new idea, as the combination of carp and ducks has long been practised in central Europe and China, where hundreds of ducks are kept on large ponds. Ducks are kept because they are fast-growing and produce a quality meat, but on their own they create a muddy pool where the high level of nutrients from their dung has the potential to cause toxic blue-green algal blooms in summer. Carp on their own never make best use of all the food resources, but carp and duck together complement each other well. Commercially the ducks are raised until they are about 50 days old and weigh about 2.5kg (5lb 8oz).

Carp ponds have long been a feature of manor houses and monasteries, where the fish were eaten on days when meat was forbidden. The carp is a hardy, fast-growing, omnivorous fish that does not need a high-quality protein pellet. In fact, carp can be quite self-sufficient and they do not demand high-quality water. They thrive in muddy pools! With carp and duck together, the ducks muddy the water and produce lots of droppings, which are a food source to the animals that live in the mud and on which the fish feed. If necessary, you can supplement the carp's diet with mealworms, which are easy to grow. Carp take a while to get to size – as much as 3 years – but the cost is minimal.

While an intensive system is not suited to a small plot, there should be enough space to incorporate a decent-sized netted pond that could support a number of carp and ducks. For example, a deep pond 10m x 4m (33' x 13') would allow you to keep a reasonable number of carp and a small flock of ducks.

A recently dug pond suitable for ducks and carp. It receives run-off from nearby roofs.

Chapter ELEVEN

Beekeeping

Honey bees are vitally important on the smallholding, as they are one of the most important pollinators. It is estimated that honey bees pollinate one-third of our food crops, and they are especially valuable if you have an orchard. Of course the other reason to keep bees is for their honey!

The bright, open flower of Elecampane attracts this bumble bee.

other bees

Honey bees are not the only bees that you will spot on your plot, as there are a number of species of bumblebees and solitary bees that will be attracted to the fruit blossom and other flowers.

Each worker bee's job is related to her age. Young workers stay in the hive, while older workers are the foragers.

Keeping bees is not as simple as going out and buying a few hens and a hen house. During the summer months a lot of time and effort goes into keeping the bees healthy and preventing them from swarming, while extracting the honey is hard work. But beekeeping certainly has its rewards: apart from the bees' value as pollinators, a productive hive can produce around 10kg (22lb) of honey a year – and there are other bee products too, such as beeswax, propolis and royal jelly.

You need to have a fair amount of knowledge to keep bees successfully. So before rushing out and buying your hive and equipment, you should attend a course about beekeeping. Even then, it is recommended that you get your bees only once you have a mentor who can guide you through the first couple of years, or you could join a local beekeeping society, who can provide support (see Resources). Membership of a club will give you access to a wealth of knowledge, updates on disease in your area and how to deal with it, and possibly some free insurance and loan of specialist kit.

If you can't spare the time that is needed for beekeeping each week, then another option is to allow beekeepers to keep their hives on your holding. That way you can enjoy the bees and their benefits without all the hard work – and get some honey as rent!

The following is just a brief introduction to the extensive subject of beekeeping, outlining the life cycle of bees and the type of kit you will need to buy in order to get started.

social organization

The social structure of the colony depends on a complex system of communication, both visual and chemical. Hormone-like chemicals called pheromones are released by the queen to unify the colony and give it an identity. She releases a specific blend of pheromones called the 'queen signal', which maintains her supreme position, suppressing the rearing of any other queen and preventing worker bees from reproducing. When she dies or weakens, the loss of this signal results in workers raising some new queens.

The workers communicate the location of food sources through the 'waggle dance', which indicates the direction and distance other workers should fly when they leave the hive on a foraging trip.

A typical National hive in spring.

The colony

A colony of honey bees typically numbers up to 60,000 bees. It is made up of a queen bee, a thousand or so drones, and the rest are worker bees.

Workers

Worker bees are female. They live for about 6 weeks and have various jobs in the colony. Each worker's job is related to her age. As young workers they stay in the hive, nursing the larvae, keeping the hive clean, building and repairing the honeycomb and regulating the temperature of the hive. If it's too hot, they beat their wings to ventilate the hive, while in the colder months, they will huddle together to generate heat.

The older workers are the foragers. They leave the hive to gather nectar and pollen, which is brought back to the hive in special pollen baskets on their hind legs. They will fly more than 5km (over 3 miles) from the hive in their search for food. When they return, they hand over the pollen and nectar to other workers. Honey is made by evaporating the water from nectar and is then stored in cells within the wax comb. Each comb is made up of hundreds of these cells, which are hexagonal in shape.

Drones

Drones are males. They are large bees with a distinctive head and long wings that reach to the end of their abdomen. They lack a sting. Drones are cared for by the worker bees and their only job in the colony is to fertilize the queen, then they die. Any drones still present in the hive at the end of summer are forced out by the workers.

This queen bee, in the centre, has been marked by the beekeeper so she is easily spotted on the frame.

The queen

The most important bee in the colony is the queen. She is a large bee, about three times the size of a worker, with long wings and a small head. She can live for several years. The queen is fed, cared for and protected by the worker bees, and her only role is to reproduce. The queen controls the colony through the release of chemicals called pheromones.

When an old queen weakens, the workers start the process of replacing her, which is called supersedure. A few newly hatched larvae are picked to become a queen, and these are fed royal jelly. Royal jelly is a white substance made by the workers from pollen and glandular secretions rich in nutrients and vitamins, and it fuels the larvae's growth. Supersedure will also take place if the queen dies unexpectedly.

The new queens fight among themselves and eventually only one remains. She then leaves the hive with the drones and goes on her nuptial (mating) flight. She mates with several drones during the flight, and in the process accumulates sufficient sperm to fertilize eggs for the rest of her life. Once she has mated she returns to the hive, where she lays up to 2,000 eggs a day, one egg per cell. Each egg is the size of a grain of rice. Fertilized eggs become workers and occasionally queens, while unfertilized eggs hatch into drones.

The eggs hatch into larvae, or grubs. The larvae remain in their cells, where they are fed and cared for by the worker bees. The cells are sealed by the workers once the larvae are ready to pupate and metamorphose into adults. A new worker bee emerges after 21 days, a drone after 24 days and a new queen after just 16 days.

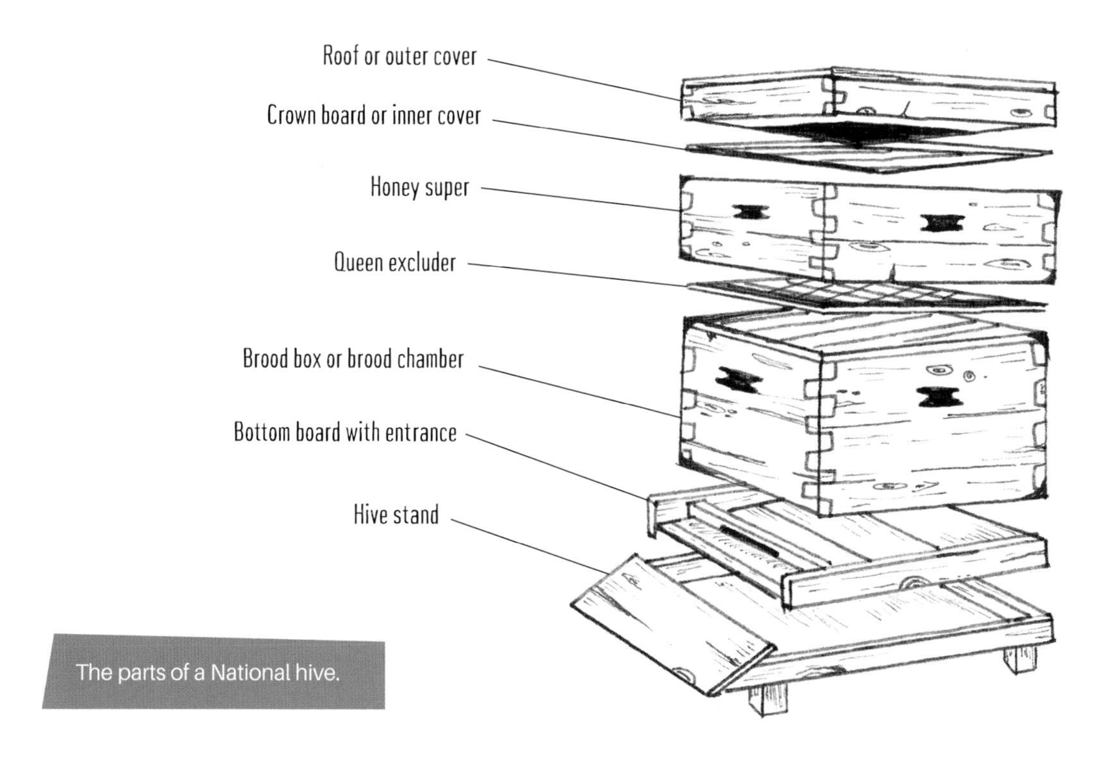

The parts of a National hive.

Inside a hive

In the wild, a colony of honey bees would make its nest in hollow trees and cavities in buildings. A hive is similar, in that it's a structure that is hollow, but has been designed to make the management of the colony easier. Around the world there are a number of hive designs – for example the Langstroth, popular in Australia and North America; the WBC, with its classic hive shape; and the National, which is a favourite in the UK. However, they all have the key parts in common, namely the roof or outer cover, crown board or inner cover, honey supers, queen excluder, brood box or chamber, bottom board and stand.

Inside the brood box and supers are frames, which hang vertically inside them. Each frame has a wax sheet that the bees use as a foundation layer on which to build their wax comb.

The colony lives in the brood box, which is where the queen lays her eggs. Some of the cells are filled with pollen and honey and provide an easily accessible food source for the queen, drones and workers.

The supers and the frames within them are usually shallower than the brood box. It is here that the worker bees store honey. They fill the cells with honey and add a cap of wax to keep it in. It is the honey in the supers that the beekeeper harvests. You will need a number of supers. In late spring, when the 'flow' starts, the frames in the first super will quickly fill up, so you need to keep adding extra ones on top to provide more space for honey storage. A queen excluder is

In this apiary, the two hives on the left have large colonies and contain a number of supers full of honey.

Bees have regular flight paths, so put up screens if you don't want them flying low across your main paths.

placed between the brood box and the honey supers, so the queen bee does not crawl up into the super and start laying eggs.

The bottom board is usually placed on a stand so that the hive is off the ground and easier to handle. Many beekeepers opt for a screened bottom board with a mesh floor, which provides good ventilation for the hive – something that is essential for healthy bees. It also allows any debris from the cells to drop out of the hive. The front of the bottom board is cut away to form the entrance for the bees. The roof or outer cover provides protection from the rain and helps to insulate the hive in winter.

Essential kit

As well as a hive, you will need some protection in the form of a bee suit with veil and gloves. You can opt either for a veil with a jacket or, for greater protection, a veil with a full suit. You can buy leather gloves, but cheap rubber gloves are just as good. You will also need a smoker, which, as the name suggests, produces smoke that is used to calm the bees when you inspect the hive. Rolled-up cardboard or chips of wood are burnt inside the smoker, and the smoke can be puffed into the hive entrance. You will also need a hive tool: a flat metal bar that you use to prise apart bits of the hive. Bees stick everything together with propolis, making it difficult to lift off parts of the hive, hence the tool. Other extras include a bee brush or a large goose feather, to brush bees off a frame, a toolbox in which to place all your bits and pieces, a marking kit and possibly a queen catcher.

There is one other piece of equipment that you may need, and that's an extractor. It's an expensive piece of kit, used to extract the honey from the frames at the end of summer. You *can* extract the honey manually, but it is a time-consuming and very messy job. The extractor spins the frames and the honey drains out and is collected at the bottom. There are cheaper plastic extractors available online, but if you belong to a local beekeeping association you may be able to hire a good-quality one.

Siting your hive

Ideally, your hive should be placed in an area that is not cultivated, so there is less disturbance on a daily basis, such as a corner of a pen not used for livestock or cultivation. But don't forget that a honey-filled super is very heavy, so you want to be able to reach the hive with a wheelbarrow or trolley. You will need a bit of space around the hive to be able to carry out your inspections, and maybe to expand the number of hives. Bees have regular flight paths, so if you don't want them flying across your main paths, you can force them to fly higher by putting up screens, or planting hedges.

Getting your bees

Once your hive is set up you can start to look for your new colony. Often you can buy an entire colony from an established beekeeper who is looking to get rid of a surplus hive. Since a colony is typically 50,000-60,000 bees, this can be quite daunting for a novice. Alternatively, you can buy what is called a nucleus or 'nuc'. This is a small colony of about 10,000 bees, with a young laying queen and some brood on five frames. A nuc is the easiest way to get started, as the number of bees is small and manageable, giving you time

Bee conservation

Around the world, bees are under threat. Not only are they affected by harmful parasites such as the varroa mite, but there are new pesticides that present a deadly risk. This group of pesticides, called neonicotinoids, is used to protect crops such as oilseed rape from the flea beetle. They are systemic pesticides, used to treat the seeds, so the whole plant is impregnated with the chemical, including the pollen. The use of this class of pesticides is now known to be having a devastating effect on bees. They are neurotoxins and cause the bees to be disoriented and colonies to fail. Both neonicotinoids and the varroa mite are associated with the mysterious and alarming 'colony collapse disorder' – the unexplained disappearance of whole colonies of bees across North America and Europe. Neonicotinoids are now banned in some countries, but sadly are still in use elsewhere.

A swarm of honey bees clinging to a tree.

to get used to them before handling larger numbers. The third way is to buy a swarm, but this is really down to luck! Local beekeepers may be called to collect a swarm, and they will give or sell on the colony to somebody wanting to get started. The downside of starting with a swarm is that you know nothing about the bees – such as where they came from, the health status of the colony and their temperament – and yes, you do get bees with a nasty temperament!

Natural beekeeping

There is an alternative to the traditional way to keep bees, and that is natural beekeeping, which makes use of simple hive designs that have been around for thousands of years and are still used in much of the developing world. It is a less intensive way of beekeeping, and the bees are kept mostly for the benefit they bring rather than for a harvest of honey. It's based on the approach of 'giving to the bees', rather than 'taking from the bees'.

Most natural beekeepers use a top-bar hive, which is easy to make. It is simply a trapezoidal box with bars across the top and an entrance for the bees. Each bar has a groove along its length that is filled with wax to act as a guide, and an anchor point for the bees to build their comb. One of the advantages of a top-bar hive is the ease of managing it. It's very much a hands-off style of beekeeping, with minimal disturbance.

It is possible to take a small harvest of comb honey from these hives, but the bulk of the honey is left for the bees so they have enough

The hive on the left is a Warre hive, which looks like a traditional hive but is in fact a natural hive, while the white-and-blue ones are top-bar hives.

Natural beekeeping is based on the approach of 'giving to the bees' rather than 'taking from the bees'.

food to last the winter. This avoids the need to feed the bees with a large quantity of sugar solution (tens of kilos) at the end of summer to replace their honey.

The bees benefit too. In the wild, the size of the cell varies to suit the conditions. In a traditional hive, the wax foundation dictates the size of the cell, which tends to be larger than a wild cell, so there is more room for honey. But a larger cell also encourages the varroa mite, a major parasite of bees and responsible for the failure of many colonies. In a top-bar hive the bees determine the size of the cells, and there seems to be a lower incidence of varroa in these hives.

Extraction is simple. First, you remove a bar with its comb, brush off the bees and replace it with a new bar. Drop the comb into a stainless-steel bucket and crush it to release the honey. Strain the honey through muslin into a jar. The leftover wax and honey can be placed back near the hive for the bees to clean, and then you can use the wax for candles.

APPENDIX: Livestock regulations & good practice*

Animal welfare

Good animal welfare is key, and for many people one of the reasons for keeping livestock is to ensure that the eggs and meat they eat come from animals that have had a good life. It is important to know your animals, so that you can quickly spot problems. Check them twice a day and take time to observe them. Watch when they run up to you when you bring food: are they moving OK; are they lame; are they behaving differently? Are their faeces normal; is the water going down more quickly than usual? With chickens you could do a regular feather check, as feather pecking is a sign that things are going wrong: there may be parasites or overcrowding, or the chickens are just bored and need some stimulation in their pens. The FeatherWel website (www.featherwel.org) has plenty of information on this.

Make sure you are registered with a vet and ask for an introductory visit so that he or she can check your animals and discuss welfare. Being registered also means that the vet can prescribe certain drugs, such as antibiotics, without having to visit first.

It is useful to have a medicine cabinet with essential items such as a thermometer, syringes and needles, scissors, antiseptic spray, wound powder, disposable plastic gloves, Hibiscrub (skin cleanser), foot spray for sheep and goats, a disinfectant for cleaning housing and equipment, petroleum jelly (Vaseline), a red-mite dust containing diatomaceous earth, poultry tonic, apple cider vinegar (a pick-me-up for poultry), anti-feather-pecking spray, poultry comb-and-wattle protector (cream for use in cold weather to stop frostbite) and nutri-drops (a pick-me-up for poorly lambs or chickens).

Food waste

Nowadays, animal products are moved all around the world, both legally and illegally, so there is a huge potential for these products to carry disease-causing organisms, especially viruses – which may not be killed by freezing or curing. Diseases that can be contracted from eating contaminated food include foot-and-mouth disease, African swine fever, Aujeszky's disease and swine vesicular disease. As a result, most countries have strict rules regarding the feeding of food waste to livestock, especially meat-containing waste.

Some countries, such as Australia, have restrictions only on the feeding of meat-containing products, but in the EU there are rules that ban the feeding of any food waste from a catering establishment or domestic kitchen to livestock, including chickens. This includes all kitchen scraps even if they are vegetable-based, as there is a risk that they may have come into contact with meat. I find it's much easier to pick over my fruit and tidy up vegetables while I'm on the plot, so the waste can go straight to the animals or in the compost bins, rather than deal with it in my kitchen. Anything coming out of the kitchen goes to the wormery or the compost bin.

Rules and regulations

There are many rules and regulations relating to the keeping of livestock, in particular to prevent the spread of disease, and these may include identification requirements and movement restrictions. The rules apply whether you keep hundreds of animals or just a couple. They vary from

*This section is available as a pdf with live hyperlinks at: www.greenbooks.co.uk/one-acre-livestock-appendix

country to country and even state to state, and they change regularly, so it would be impossible to describe them all in the space available here, but the following is an outline of the scenarios in the UK, USA and Australia.

For UK readers

UK-based readers can find detailed guidance on the regulations relating to the keeping of farmed animals on the websites of the Department for Environment, Food & Rural Affairs (Defra) (www.defra.gov.uk) and the Food Standards Agency (www.food.gov.uk). Don't forget that regulations vary between England, Scotland, Wales and Northern Ireland!

Holding number

If you keep a few chickens or only grow fruit and vegetables then you have no need to register your holding. But if you keep livestock such as goats, sheep, pigs or cattle, or wish to apply for agricultural grants or subsidies, you must obtain a County Parish Holding (CPH) number. This is issued by the Rural Payments Agency (www.gov.uk/rural-payments). The CPH is a 9-digit number: the first 2 digits relate to the county, the next 3 relate to the parish and the last 4 digits are a number unique to the keeper (e.g. 12/345/6789).

You will also require a flock or herd number, which is obtained from your local Animal Health Divisional Office (AHDO). The number is used to identify your stock and is linked to your CPH. The AHDO will send you a registration document, which will contain your personal details, CPH number and herd mark. If you keep a total of more than 50 birds of any species, you need to register your poultry with the Animal and Plant Health Agency.

Keeping records

Your local AHDO will issue you with record books in which you must record all movements, births and deaths, so at any one time you know how many animals you have on your holding, and where animals have been moved to or from. An animal health record book is used to record the use of any medicines, the place from which you obtained the medicine, how much was given and the withdrawal period (the length of time that must elapse before the treated animal may enter the food chain, allowing the medicine to break down in the body so there are no traces of it in the meat). Most counties require you to submit your records every few years. You are not obliged to use the issued record books, so you can use the various farm management packages and apps instead, but you must be able to print out a report to send in when requested.

Animal identification

All stock needs to be identified. In England, sheep and goats must be identified within 9 months of birth if they are not housed at night, or 6 months if they are, and before you move them off their holding of birth if that is sooner. Adult sheep must have two tags bearing the same unique number, and one must be a yellow electronic identifier (EID). Animals that are sent to slaughter before they are 12 months old can be identified with a single EID ear tag. There is no requirement to read and record the EID number in the tag, but it will be possible for the abattoir to do so. Pigs need to be identified before they are moved, by either a slapmark (a permanent ink mark on each shoulder), an ear tag or a tattoo in the ear with your herd mark, or, in the case of weaners, by a temporary paint mark. You can order tags from approved manufacturers, together with the applicator needed to insert the tag. Goats must be tagged, but are not yet required to have an EID tag. Rules in Scotland, Wales and Northern Ireland are different.

Movement regulations

For sheep, goats and pigs you must have a CPH number before moving stock on to your holding, and you need to report the movement. You must also report movements off the holding, to another holding or to a slaughterhouse – a process that can now be done electronically – and you must also record all movements on and off your holding in your holding register. When animals arrive on your holding there will be a standstill period, when livestock cannot be moved off your holding, to help prevent the spread of disease. For example, when you move a pig on to a holding, no pig may be moved off it for 20 days other than to slaughter, and no cattle, sheep or goats can be moved for 6 days.

Fallen stock

Dead animals need to be collected or taken to an approved site for incineration. They must not be buried on the farm. When you have livestock on your holding you should find a local collector whom you can contact in the event that you have dead animals to dispose of. It is important to keep the receipt given to you by the collector so you can prove that your dead stock were disposed of in a correct manner. A list of approved sites can be found online, but it is easier to use the services of the National Fallen Stock Company (www.nfsco.co.uk).

Selling eggs

There are various rules that apply to the selling of eggs. Mostly chicken keepers have 'farm gate' sales, for which eggs must be ungraded, clean but not washed, undamaged and with a 'best before' date that is within 4 weeks of laying.

If you intend to sell eggs and have more than 350 birds, you need to register with the Animal and Plant Health Agency (APHA) (www.gov.uk/government/organisations/animal-and-plant-health-agency). You must also register if you sell any eggs to a packing centre or you have more than 50 hens and sell your eggs at market. If you wish to sell eggs to shops or catering establishments, you must be approved and authorized as a packing centre by the APHA so you can grade the eggs as Class A. You are given a code, which must be stamped on all Class A eggs. Producers are required to keep detailed records relating to their birds. If you have fewer than 50 hens, you can still sell eggs at a local market, but you need to display your name and address, the 'best before' date and any advice to consumers as required under the food labelling regulations.

If you are transporting ungraded eggs from your holding to a packing centre or a catering establishment, you must register as a food business operator with your local authority and retain records of every consignment of eggs.

Aquaponics

If you intend to keep non-native fish, you must register with the Centre for Environment, Fisheries and Aquaculture Science (CEFAS) and complete an ILFA1 form, and if you intend to trade live fish or run an educational centre you will need to have an inspection and complete Form AW1. Full details can be obtained from CEFAS (www.gov.uk/government/organisations/centre-for-environment-fisheries-and-aquaculture-science).

Slaughter regulations

EU regulations allow for the on-farm slaughter of small numbers of birds, which can be sold from the farm gate, to local shops, at markets in your own county and adjoining counties, or direct to mail-order customers. If you intend to slaughter poultry on your holding and sell the meat, you need to be registered with your local Environmental Health Department and you must hold a licence to stun and kill poultry. The birds must be stunned and rendered unconscious by electrical or gas stunning or by captive bolt, and then killed immediately by bleeding (cutting the neck arteries). Guidance regarding the correct method of stunning and killing poultry can be obtained from the Humane Slaughter Association (www.hsa.org.uk).

Sheep, goats and pigs should be taken to a local legally approved abattoir for private slaughter. Although the regulations do allow for the home slaughter of sheep, goats and pigs by the animals' owner, there are a number of requirements to be met in order to safeguard animal welfare and food safety. If you have arranged for an animal to be slaughtered on your holding, you must ensure that the animal does not suffer any unnecessary suffering or distress. Also, the meat from the slaughtered animal may be eaten only by you and your immediate family. It is an offence to sell or supply meat that has not been slaughtered and health-marked in a licensed abattoir. See www.gov.uk/guidance/meat-and-meat-hygiene#home-slaughter-of-livestock.

Planning

If you have purchased some land and are intending to erect buildings such as a stable, shed or polytunnel, you will need to check with your local authority to make sure you do not need to apply for planning permission. Also, you will require planning permission if you want to make a new entrance off the road.

For US readers

The complex nature of regulations at the levels of state and federal government mean that it is not possible to go into any detailed discussion here. Livestock traceability is just as important in the USA as in other parts of the world, and there are requirements to obtain premises ID numbers and ear tags for livestock. I would advise that before buying any livestock you refer to your state body or the USDA website (www.usda.gov) and the Animal and Plant Health Inspection Service (APHIS) (www.aphis.usda.gov). There are also rules regarding the keeping of animals in urban and rural areas, which may restrict the numbers of animals that may be kept. These tend to vary from county to county, and you may be required to obtain a permit.

For Australian readers

All properties used for agricultural purposes are required to have a Property Identification Code (PIC), which is an 8-character alphanumeric code issued by the state or territory authority. The first numeral represents the state/territory, which is followed by 2 letters that represent the municipality, then 2 more letters and 3 numerals. You must apply for a PIC if you keep cattle, sheep, goats, pigs, alpacas, llamas, deer, horses or more than 100 poultry, and, in some states, have 0.5ha (1¼ acres) of grapevines or 20 or more chestnut trees.

Australia has a National Livestock Identification System (NLIS) in order that cattle, sheep, goats and pigs can be tracked during disease outbreaks. All livestock must be identified with an approved NLIS tag before being moved off a property. If the animal is being moved off its property of birth, it should bear an NLIS breeder tag, which remains with the animal for life. If this is lost, it is replaced by an NLIS post-breeder tag. The colour of the breeder tag represents the year of birth, while all post-breeder tags are pink. At the time of writing, there is an exemption to tag certain breeds of dairy goats. All movements of livestock are accompanied by a National Vendor Declaration (NVD), showing where they have come from and where they are going.

Owners of more than two pigs require a pig tattoo brand (owners of no more than two are exempt). The brand starts with a numeral that indicates the state, for example Victoria is 3. All pigs going to sale or slaughter must be identified by either a tattoo brand or an ear tag, depending on their body weight, before they leave their property of birth. Pigs under 25kg (55lb) must be tagged with an approved NLIS ear tag printed with the numbers of the tattoo brand. Heavier pigs need both an ear tag and a tattoo brand. Any movement must be accompanied by a PigPass NVD, which contains information to meet traceability and food safety requirements.

Beekeepers too have to register, and brand all their hives with their unique identifier as proof of ownership. They must report any outbreak of disease, including American foulbrood (AFB), parasitic mites, small hive beetle (SHB) and the bee louse, to their local Department of Agriculture.

Information about Australian identification and movement regulations can be found on the Meat & Livestock Australia (www.mla.com.au) and the National Livestock Identification System (https://nlis.com.au) websites.

RESOURCES*

A wealth of information is available on the topics covered in this book, either in printed form or online. Here I have highlighted some of the resources that may prove to be most useful.

Informative websites

Agriculture and Horticulture Development Board (AHDB)
www.ahdb.org.uk
UK levy board providing information on beef and lamb, pork, potatoes, cereals and horticulture. The sheep and grassland manuals of the Better Return Programme provide essential reading on pasture management.

Agroforestry Research Trust
www.agroforestry.co.uk
UK charity that researches into temperate agroforestry, including tree, shrub and perennial crops. The Trust Director is Martin Crawford *(see page 217)*.

Animal Welfare Approved
http://animalwelfareapproved.org
US-based organization providing the certification for the Animal Welfare Approved label. The website includes useful information regarding welfare standards.

The Aquaponic Source
www.theaquaponicsource.com
A comprehensive website selling equipment, with extensive online information regarding the setting up and management of aquaponic systems. Based in the USA, but a valuable resource for would-be fish keepers around the world.

Australian Pig Breeders Association (APBA)
http://lbcentre.com.au/Australian_Pig_Breeders_Society_Australia.php
The APBA manages the herd book for nine breeds of pigs and promotes the pedigree pig.

Backyard Aquaponics
www.backyardaquaponics.com
Essential resource if you are considering a small aquaponics system of your own. Based in Australia.

British Beekeepers Association
www.bbka.org.uk
National organization supporting beekeepers and working to conserve bees.

British Pig Association (BPA)
www.britishpigs.org.uk
The BPA maintains the herd book for most rare pig breeds, with information on the breeds and buying pedigree pigs.

Cotswold Grass Seeds
www.cotswoldseeds.com
This is a commercial website, based in the UK, but it includes some really valuable web pages giving information on grass mixes, green manures and establishing nectar beds, with links to some excellent articles.

Empire Farm
www.empirefarm.co.uk
My own website, featuring the one-acre plot, with a blog of activities that take place on the plot during the year.

Four Season Farm
www.fourseasonfarm.com
The experimental organic market garden in Maine, USA, owned by Barbara Damrosch and Eliot Coleman.

Garden Organic
www.gardenorganic.org.uk
A UK campaigning and research charity. The website is full of information on all organic horticultural matters.

Haller Foundation
www.haller.org.uk
Charity promoting sustainable agriculture in Africa, with particular emphasis on aquaponics.

Holderread Farm
www.holderreadfarm.com
The Holderread Waterfowl Farm and Preservation Center in Oregon, USA, is home to the well-known waterfowl expert Dave Holderread. Website includes information about pure-bred waterfowl.

Homestead
www.homestead.org
A library of short articles on a wide range of topics. Based in the USA.

*This section is available as a pdf with live hyperlinks at: www.greenbooks.co.uk/one-acre-resources

Humane Slaughter Association
www.hsa.org.uk
UK charity. An essential website if you are planning to slaughter your own poultry or are considering the home slaughter of sheep, goats or pigs, with an online guide to the slaughter of poultry, including the use of electric stunners.

Kentish Cobnuts Association
http://kentishcobnutsassociation.org.uk
Plenty of information available as downloads if you intend to grow cobnuts.

My Pet Chicken
www.mypetchicken.com
A useful website for the first-time chicken keeper, with some guides and FAQs, plus lists of suppliers. Based in the USA.

Permaculture Association
www.permaculture.org.uk
UK charity supporting the learning and use of permaculture.

Permaculture Institute USA
www.permaculture.org
US non-profit organization dedicated to the promotion of permaculture.

Permaculture Research Institute
http://permaculturenews.org
Australian organization working with individuals and communities worldwide to expand the knowledge and use of permaculture.

Polyface Farms
www.polyfacefarms.com
Polyface Farm in Virginia, USA is the home of Joel Salatin, a low-carbon farming expert who lectures around the world on sustainable farming. His videos are essential viewing, especially those on his poultry chicken tractors.

Poultry Keeper
https://poultrykeeper.com
One of the most comprehensive websites on keeping poultry of all kinds, with a useful forum. Based in the UK.

Rodale Institute
http://rodaleinstitute.org
Founded by organic pioneer J. I. Rodale, to study the link between healthy soil, healthy food and healthy people. Useful information on organic management, with webinars and online courses. Based in the USA.

Soil Association
www.soilassociation.org
The UK's leading membership charity campaigning for healthy, humane and sustainable food, farming and land use. The website has plenty of information and downloadable pdfs on the organic management of farms.

Soil Foodweb
www.soilfoodweb.com
Website covering the work of the soil expert Dr Elaine Ingham. Based in the USA, but with many online resources and webinars.

Magazines

Subscribing to a magazine can really boost your knowledge. It can also be useful to dip into magazines published in countries other than your own, for new approaches. You can buy one-off editions of some of the magazines listed below via digital newsstands or by contacting the publisher directly.

Acres USA
Published monthly in the USA (Austin, TX).
www.acresusa.com
A leading publication covering organic and sustainable farming. Aimed at production-scale farms, many of the ideas can be applied to small-scale operations.

Country Smallholding
Published monthly in the UK (Barnstaple, Devon).
www.countrysmallholding.com
The UK's leading monthly for smallholders, covering a range of topics.

Grow Your Own
Published monthly in the UK (Colchester, Essex).
www.growfruitandveg.co.uk
Comprehensive coverage of topics relating to fruit and vegetable growing, plus flowers and chickens.

HomeFarmer
Published monthly in the UK (Preston, Lancashire).
https://homefarmer.co.uk
The focus of this magazine is home-grown and home-made, with a lot of DIY ideas.

Kitchen Garden
Published monthly in the UK (Horncastle, Kent).
www.kitchengarden.co.uk
Down-to-earth advice on growing fruit and vegetables.

Permaculture: Practical solutions for self-reliance
Published quarterly in the UK (East Meon, Hampshire).
www.permaculture.co.uk
A variety of topics on the theme of permaculture from around the world.

Practical Pigs
Published quarterly in the UK (Kelsey Media, Cudham, Kent) in collaboration with the British Pig Association (BPA).
www.kelsey.co.uk/pigs
A great magazine for pig keepers, with lots of practical advice and guidance on breeds of pigs.

Practical Poultry
Published monthly in the UK (Kelsey Media, Cudham, Kent).
www.practicalpoultry.com
Informative and with lots of ideas, from incubation to showing and selling. Good advice on housing and food.

Small Farmer's Journal
Published quarterly in the USA (Cedar Sisters, OR).
https://smallfarmersjournal.com
Aimed at the homesteader and small farmer, the magazine covers lots of useful topics.

Small Farms
Published monthly in Australia (Bowral, NSW).
www.smallfarms.net
For the small farm owner, especially those entering the farming sector for the first time.

Smallholder
Published monthly in the UK (Falmouth, Cornwall).
www.smallholder.co.uk
A magazine for small producers and for the self-reliant household.

Books

There is a vast range of books providing expert advice on all the topics covered in this book, of which a small selection is included here. In addition, I would recommend reading some of the classic books on soil and organic agriculture dating back to the 1940s and 1950s, especially those written by Sir Albert Howard, as well as by Lady Eve Balfour, Louis Bromfield, Jerome I. Rodale and Rudolf Steiner: all pioneers of the organic movement. Most are no longer in print, but may be found in second-hand bookshops and via specialist websites.

Design and groundwork

Gaia's Garden: A guide to home-scale permaculture
Toby Hemenway (Chelsea Green Publishing, 2nd edition 2009)
A readable introduction to permaculture, helping you create an ecological home garden.

The New Organic Grower: A master's manual of tools and techniques for the home and market gardener
Eliot Coleman (Chelsea Green Publishing, 2nd edition 1995)
If you aspire to grow organic vegetables, this may be the book for you: with everything about organic vegetable growing – from fertility and crop rotations to materials, costs and even marketing.

The One-Straw Revolution
Masanobu Fukuoka (New York Review Books Classics, 2009; first published 1978)
A revolutionary approach to sustainable agriculture, 'natural farming', developed by the author in Japan.

The Polytunnel Handbook
Andy McKee and Mark Gatter (Green Books, 2008)
A manual that covers all aspects of polytunnel ownership, from planning and building to cropping.

Sepp Holzer's Permaculture: A practical guide for farms, orchards and gardens
Sepp Holzer (Permanent Publications, 2010)
The 'rebel' Austrian farmer writes about his natural approach to farming, including his use of hugelkultur and natural branch development.

Teaming with Microbes: The organic gardener's guide to the soil food web
Jeff Lowenfels and Wayne Lewis (Timber Press, revised edition 2010)
A great read in a chatty style that makes the technical stuff really easy. Full of essential information.

Growing produce

Charles Dowding's Veg Journal: Expert no-dig advice, month by month
Charles Dowding (Frances Lincoln, 2014)
A month-by-month guide to growing vegetables.

Creating a Forest Garden: Working with nature to grow edible crops
Martin Crawford (Green Books, 2010)
One of the best books on forest gardening.

Edible Perennial Gardening: Growing successful polycultures in small spaces
Anni Kelsey (Permanent Publications, 2014)
Not everyone has space for trees, but this book shows the range of edible perennials that you can grow in a small space.

The Fruit Tree Handbook
Ben Pike (Green Books, 2011)
Expert guidance on growing fruit trees, from planning and planting to pruning and harvesting.

How to Grow Perennial Vegetables: Low-maintenance, low-impact vegetable gardening
Martin Crawford (Green Books, 2012)
An inspiration to grow a much wider range of long-lived plants, for a supply of edible shoots, fruits, roots, etc.

How to Make a Forest Garden
Patrick Whitefield (Permanent Publications, 3rd edition 2012)
An introduction to creating a forest garden.

Organic Gardening: The natural no-dig way
Charles Dowding (Green Books, 3rd edition 2013)
A great introduction to the no-dig or no-till approach.

The Winter Harvest Handbook: Year-round vegetable production using deep-organic techniques and unheated greenhouses
Eliot Coleman (Chelsea Green Publishing, 2009)
Guidance on how to extend the growing season and grow vegetables through the winter months.

Keeping livestock

Aquaponic Gardening: A step-by-step guide to raising vegetables and fish together
Sylvia Bernstein (Saraband, 2013)
The best book on DIY aquaponics, based on the author's many years of experience.

The Book of Geese: A complete guide to raising the home flock
Dave Holderread (Hen House Publishing, 1993)
One of the best reference books on keeping geese.

A Guide to Traditional Pig Keeping
Carol Harris (The Good Life Press, 2009)
A good reference book for both the novice and the experienced pig keeper.

My Pet Chicken Handbook: Sensible advice and savvy answers for raising backyard chickens
Lissa Lucas and Traci Torres (Rodale Books, 2014)
Novice and experienced chicken keepers alike get something from this book, with lots of practical and common-sense advice.

Sheep: Small-scale sheep keeping
Sue Weaver (Hobby Farms, 2nd edition 2014)
Packed with information, with an easy-to-read approach.

Storey's Guide to Raising Ducks
Dave Holderread (Storey Publishing, 2nd edition 2011)
A detailed look at keeping ducks, from choosing the right breed to rearing and slaughter.

Suppliers

Most of the equipment mentioned in this book (e.g. fencing materials, feeders, water troughs, sheep hurdles, animal husbandry equipment, feedstuffs) can be obtained through agricultural merchants or through specialist websites. Poultry and pig housing, greenhouses, and polytunnel hoops and plastic are best purchased via specialist manufacturers, which can also be found online.

If you are looking to establish an orchard, grow soft fruits or plant an edible hedgerow, find a local nursery that is able to supply regional varieties suited to your growing conditions. Some of the perennial plants mentioned in Part Two of this book may be found in specialist nurseries or can be grown from seed.

Sourcing livestock can be trickier, especially if you are looking to buy a specific breed. One way to find breeders is to attend local agricultural shows and talk to the exhibitors about their animals. When I am buying animals I like to visit the breeder and assess the conditions in which animals are being raised, and only then commit to buying. If this is not possible for you, join a smallholder forum or a local smallholder group and ask questions. I find that people are more than willing to give recommendations.

INDEX

Page numbers in *italic* refer to illustrations

Also by Green Books

Creating a Forest Garden

Martin Crawford

Creating a Forest Garden tells you everything you need to know to grow edible crops, while letting nature do most of the work. Whether you want to plant a small area in your back garden or develop a larger plot, it includes practical advice on planning, design (using permaculture principles), planting and maintenance. With a detailed directory of over 500 trees, shrubs, herbaceous perennials, annuals, root crops and climbers – almost all of them edible and many very unusual – this is the definitive book on forest gardening.

Six Steps Back to the Land

Colin Tudge

Colin Tudge coined the term 'Enlightened Agriculture' to describe agriculture that is "expressly designed to provide everyone, everywhere, with food of the highest standard, nutritionally and gastronomically, without wrecking the rest of the world". In *Six Steps Back to the Land* he explains how we can achieve that, and have truly sustainable, resilient and productive farms, with successful smallholdings as one of the steps along the way.